HOBBY FARM

HOBBY FARM
Ideas for the
New Countryside

by
Willy Newlands

SOUVENIR PRESS

A writer? You must be out of your mind...
For Dorothy, with all my love.

First published in 2006
by Souvenir Press Ltd
43 Great Russell Street, London WC1B 3PD

ISBN 0285637681
9780285637685

Typeset in 11/13pt Sabon by FiSH Books, Enfield, Middx.
Printed and bound in Great Britain by
Mackays of Chatham plc, Chatham, Kent

Contents

Introduction

A Google search for the phrase 'hobby farmer' turns up half-a-million references. These modern lifestylers may be difficult to define, but they are very big business.

The number of farms in England with 'no economic output' in recent surveys jumped in one year from 90,000 to 115,000, while the number of large farms fell to only 15,000. This is more than a minor trend, it is a seismic shift in the rural way of life.

Country living is now a pastime rather than an occupation.

The Royal Institute of Chartered Surveyors (RICS) says that nearly half of all farms sold in a recent year were bought by non-farmers. This is described as 'consistent with the new trend of buying houses in the country with 20–60 acres of land – more than a garden, not quite a fully-fledged farm.'

A house with a large garden can be a mini-hobby farm, with vegetables, poultry, rabbits and a few hives of bees. An acre of land will support a couple of milking goats or some pigs, and once the hobby farmer is operating on a hectare (about 2.5 acres) she can think of poultry free-ranging in an orchard or a mini-flock of sheep.

Doubling the acreage makes it possible to keep a cow and calf, and increasing it again to 10 acres we can find space for some cropping in support of the livestock (hay and forage crops) and a really worthwhile mixed orchard with apples, plums and other fruit.

Even though each type of pleasure farm may be clearly seen by its owner as a non-profit enterprise, there is great satisfaction in making the land productive enough to cover the cost of feeding the livestock and giving some high-quality additions to the household menu, from fruit and vegetables to meat and eggs.

William Tew of the RICS says: 'These people have bought the right to live in a property with privacy and unspoilt views, but they have probably given little consideration to how they might manage their surrounding acres. Hobby farming...can be an interesting and practical solution.'

This book has been written for someone who definitely does not see himself (or frequently herself) as a mucky-handed smallholder or heavily-subsidised crofter and certainly not a barley baron. It is an upbeat review of the possibilities of having fun inside your own 'green belt'.

Hobby agriculture covers a wide spectrum, from backyard eggs-and-jam to large acreages of grazing land. The main planks on which a definition can be made are money and labour: the hobby farmer's income is largely made from off-farm work and the holding does not employ full-time labour.

There is a blurred line between the smallholder/crofter and the hobby farmer, although my own definition would be 'a smallholder tries to make money on his land, a hobby farmer spends money on his land.' Mainly, it's a matter of attitude. The noveau farmer is, above all, enjoying himself.

Artichokes: Wonderful flavour and lots of mythology

Artichokes are the subject of more misinformation and balderdash than almost any other vegetable on the menu. They can give you leprosy, cure diabetes, make you rich almost overnight, overrun your garden like knotweed, or cause such flatulence that you will be unfit to spend time in civilised company with your 'loathsome stinking winde'.

Sorting out the fact from these fictions in the world of the artichoke is not easy and you would almost expect to see a black-magic warning on the seed packet. Even farming and gardening textbooks are at odds over these interesting plants, starting with their names.

There are two artichokes – the globe (above), which is the gourmet one, with the honey-flavoured thistle buds, and the Jerusalem artichoke, which has edible tubers, again with a nuts-and-honey flavour, and which has the added virtue of making good game cover. Both are perennial and both are easy to grow virtually anywhere in lowland Britain.

The get-rich-quick stories about the artichoke are based on a pyramid-selling scheme which gripped the American Mid-West in the 1980s. The Jerusalem artichoke is a native

North American, a relative of the sunflower, and three enthusiastic salesmen developed the theory that this was God's gift to rural America. From these ' sunchoke' tubers could be made the perfect cheap bio-fuel. In only three years, their plan went from the drawing board to a multi-million dollar scam, and then collapsed. The promoters finished up in jail.

But even though the dream turned into a nightmare, the hype lingered on. There are still people trying to sell the Jerusalem artichoke as the basis for an eco-friendly bio-fuel revolution. This is a pity, because if you avoid the snake-oil salesmen, the sunchoke is a very useful plant. Its main drawback is its own over-enthusiasm, because Jerusalem artichokes spring up freely from every unharvested tuber on light land and can be difficult to eradicate if you decide you want the plot for some other crop (although a few pigs might soon do the trick). Any unwanted sunchokes are difficult to ignore, because they grow quickly up to 6ft or even 10ft in height, with strong, bristly and woody stems.

But this tough perennial can be a useful performer on difficult sites. In summer and autumn it provides leafy cover, and later it gives the pheasants something to dig for during hard weather. Planted 4–5 inches deep during April in ridges with a potato planter at double-furrow spacing, about 5ft apart, with 2ft between the tubers in the row, sunchokes will mature to provide a nutritious crop which the birds will eat very readily. The two-ounce tubers will survive through the winter in the ground without any protection although the standing crop loses its leaves after the first hard frost and the stems will blow flat unless supported by shrubs or fencing. In North America, the custom is to use the top growth for silage and lift the tubers to be marketed in winter.

For anyone unfamiliar with the Jerusalem artichoke, a simple starting point is a packet of tubers from greengrocer or supermarket. Our own starting point was a box of French 'topinambours' from a wholesale greengrocer. Pop them into the ground in the spring and see what you think of the result. If you like them, you will probably have enough tubers to plant up half an acre the next season. The safest way to store them is to leave them in the ground. They will store in this way until April or early May, protected by some wire netting if your pheasants or poultry start digging down to the tubers.

The name 'Jerusalem' probably comes from a corruption of the Italian name for sunflower, which is 'girasole', follower of the sun. The artichoke part of its name is due to the French explorer Champlain likening its flavour to the European globe artichoke. The link stuck, and *Helianthus tuberosus* found itself doubly misnamed.

There are various varieties in the vegetable salesmen's catalogues, such as mammoth French white and Stampede, but the only noticeable difference between them seems to be in the knobbliness or otherwise of the tubers. The knobbly ones give bigger and stronger plants, but the less-knobbly ones such as Fuseau are much easier to wash and prepare for the table and command a premium price in your farmers' market. Yields go 25–30 tons to the acre and it is a good plan to harvest them and re-plant each year because otherwise you will finish up with an impenetrably weedy and unproductive patch with deeply embedded tubers which the pheasants cannot reach.

The mythology which links artichokes with leprosy comes from the appearance of the tubers, which medical experts of the 1600s said looked like diseased fingers and therefore (of course) must cause the disease. And the idea that sunchokes can cure diabetes is based partly on a misspelling and partly on a half-understood fact. The starch in the tubers is in the form of inulin, which somewhere along the line became muddled up in people's minds with insulin. This starch can be processed to become fructose, a form of sugar more easily handled by diabetics, but it is certainly not a cure.

During World War II, the Jerusalem artichoke was grown widely in occupied Europe because it was an oddity outside the rationing system. Within two or three years it was one of the commonest of root vegetables. This popularity did not last. As soon as potatoes were readily available again they ousted the sunchoke from the shopping basket. In small quantities, these artichokes are a pleasantly flavoured addition to the menu. In bulk, they are difficult to digest and the descriptions of flatulence – it was John Goodyer who wrote in the early 1600s that 'they stir up and cause a filthie loathesome stinking winde with the bodie, thereby causing the belly to bee much pained and tormented' – are uncomfortably true.

Our own favourite Jerusalem artichoke recipe, avoiding the digestive problem, is for soup: 1.5lb of artichokes, large onion, 1oz butter, 0.5lb tomatoes, salt and pepper, garlic clove, 1.5 pints chicken stock. Thinly slice the tubers and chop the onion, adding both to the melted butter in a pan and sweat for 5 minutes. Skin, de-seed and halve the tomatoes, add to the pan and sweat for a further 3 minutes. Add chicken stock, crushed garlic and seasoning. Simmer for 15 minutes and reduce to puree. It tastes of honeyed chicken – delicious.

The globe artichoke is a Mediterranean native and somewhat more tender than the sunchoke as far as climate is concerned. The tastiest varieties, such as the large green Paris or Laon artichoke and the Green Globe, are at their best in the South and Midlands. Farther north, the Purple Globe is more reliable. And everywhere this thistly 5ft plant benefits from rich and well-manured soil.

One of my old gardening books starts off its chapter about globe artichokes with the warning that they are 'an acquired taste'. The British have acquired the taste in a big way – this is one of our most approved starters at the dinner table, giving the opportunity for lots of fiddling about, dipping in vinaigrette, and a tiny reward of tasty flesh at the base of each bract around the flower bud, plus (under the inedible, bristly choke) the delicious heart.

These artichokes are not highly productive like the Jerusalem version, nor very long-lived, and very few hobby farmers would attempt to grow them for profit. Starting from seed, they will not be productive until they are two years old, although young plants reared from offsets might produce a few usable buds in their first summer. They seem to run out of steam within 3–4 years, at which point they must be replaced, so this is definitely a luxury plant. Globe artichokes look very fine in the garden, but you have to be satisfied with half a dozen usable heads from each one, and the season is short.

Grow your gourmet plants in groups of three, each one about two handspans from the other two, and each group four feet from the next. Once in production, the king head, the primary bud, will be ready for harvesting in July, with perhaps five or six secondary buds, 3–4 inches in diameter, maturing two or three weeks later.

And if you want to grow globe artichokes just because they look good as a backing for your herbaceous border, you would be better off with cardoons. These are (probably) the wild ancestor of the globe artichoke and their silver-grey leaf is even more magnificent.

Asparagus: Take your time, it's worth the wait

British asparagus is the finest in the world. Gourmets are widely agreed on that. Food of the gods, they say. And to grow it you need the patience of a saint.

To enjoy a productive bed of asparagus on your smallholding, you have to think in timescales more appropriate to fruit trees than root vegetables. From the time you turn the first spadeful of your asparagus bed to the day you pick the first delicious spears, more than two-and-a-half years could have passed. It's a true luxury but it is very slow.

The good news is that once you've established your asparagus, it will go on producing a crop every April–June for 25 years or more. At least one plant is known to have given a harvest for 120 years. And the crop comes early, before any other luxury vegetable from the garden.

The timescale looks like this: in the early winter of the first year, you mark out a bed five paces square, digging deeply because the asparagus crowns – which are one-year-old bunches of fat roots with a few buds on top – not only need a space of nearly 4ft between rows and a foot between the plants, but will eventually go down 10ft or more in search of nourishment. Deep digging gives them a good start. In heavy soils, a raised bed is helpful.

Order the crowns, which cost nearly £1 each, during the winter and plant them in shallow trenches in the bed in March, waiting for a few spears to appear later in the spring. No picking is allowed in their first season. When these shoots come through the ground, they look like the familiar asparagus of the greengrocer's shop. Later they grow to 6ft in height, expanding into feathery sprays of foliage.

In the following winter, the plants will be knocked down by the frost and nothing reappears until the spring. When the young shoots come up, they look quite vigorous and the temptation is to start cutting the spears of these year-old plants. However, most experts again urge caution: one of the commonest causes of failure with asparagus is giving in to the temptation to take a crop too early. A handful of spears is the most you can promise yourself. Even the most enthusiastic of gardening catalogues will guarantee no more than 'a few' at this point.

After their second winter, the crowns will come into production, but even then their season will be short. There is a two-four-eight rule which is followed by successful asparagus growers. In the their third year in the bed, plants have a two-week picking season. After that, later shoots are left to develop and grow to their full height. In the following year, they have a four-week season and in their fifth and subsequent years they will produce for the full eight-week season. With most varieties in Britain this can be expected to run from late April through until June. A 40ft row of mature asparagus gives about 12–20lb of delicious spears for the table.

Top TV chefs are unanimous that British asparagus is

worth waiting for. In late winter, our supermarket asparagus comes mainly from Spain. After June, it comes from Peru. In both cases, it is a high-mileage vegetable, and the spears are not great travellers. They are at their best when freshly picked – tastier and much less fibrous.

Delia Smith says: 'English asparagus is without any doubt the best in the world. Asparagus has a painfully short season in England – just two months, May and June. So we all need to be on full asparagus alert and make absolutely sure we feast appropriately and not let the season whiz by.'

Jamie Oliver says: 'When good British asparagus is in season there is absolutely nothing like it. Simply cooked, boiled, steamed, grilled or roasted, it can be served as a starter or as a side dish and served with cooked meat or fish. When buying asparagus, I always look for plump, deep green asparagus, but the thin ones can be just as gorgeous. I don't believe in mucking around too much with asparagus – whatever flavours you add should be subtle, such as good olive oil or butter; or extremely complementary, such as anchovy butter or gorgonzola cheese.'

And Gary Rhodes adds to the chorus of approval: 'I love to make use of the best British ingredients that each season has to offer. As May arrives it brings the freshest and finest asparagus in the world. The new purple green spears can be used in all manner of recipes but actually need little more adornment than melted butter and seasoning. Simply take advantage and use while you can – remembering the season is so short and will be over by the end of June!'

Asparagus can even claim to be a guide to wealth and social status. It is said in America that you can gauge a person's wealth by how far up he or she cuts off the tip. Low income, low cut. High income, high cut.

Out in the garden, once the picking season is over and the

surviving shoots are allowed to grow tall, the ferny foliage of asparagus has its own charm. A 5–6ft screen of the lacy leaves looks good against a south-facing fence as a background to a herbaceous border, although you do have to make sure that the asparagus is not shaded out because it demands a long growing season and plenty of sunshine. Energy is stored in the spreading network of thick roots and these power the fast emergence of the spears in the spring. Plants will grow in shadier spots and thinner soils – however, the crop is hardly worth picking.

Asparagus naturally produces male and female flowers on separate plants, but several of the varieties now sold by nurserymen are all-male (Franklim, the best seller, and Frolim are examples). These do not produce the bright red berries of the female plant in late summer. This is said to help the plants to maintain their vigour. The male spears are more abundant but slightly thinner-stemmed than female shoots.

Most commercial crops of asparagus in this country are grown in Norfolk, Suffolk, Cambridgeshire and Lincolnshire, with outposts in Essex, Kent and the Vale of Evesham. There are half a dozen major producers who are extending the season by using polytunnels and constantly upgrading the varieties used in their attempts to win more of a market which is currently estimated to be worth £50 million a year. Only one third of this is supplied by British growers and the rest is imported.

Market watchers say there is 'huge potential for expansion,' a phrase which you do not hear very often in connection with any aspect of English agriculture. Smallholders can aim for farm gate sales and also for the market in yearling crowns. These retail at about £1 each. With seeds of named varieties costing less than a penny each, the possibilities of breaking into this business might also be worth exploring. Asparagus is usually germinated in pots under glass and planted out in summer to grow on to saleable size.

Asparagus is quite an adaptable plant in some ways, particularly in its tolerance of salt in the soil. Its native home is in the salt-marsh estuaries of the eastern Mediterranean and it doesn't mind an occasional splash of seawater. It also withstands high temperatures – up to 100°F plus in California – and bitter winters with the thermometer down to 40° below freezing (on both the C & F scales).

In California, the luxury vegetable is mega-business. Perhaps Britain's growers should make a serious effort to copy the Stockton Asparagus Festival, which is estimated to be worth $19 million to the district and is rated as one of the '100 Best Events in the Nation' by Destination magazine. The festival is one of the 'Seven Best Events in the State' according to the California Department of Tourism.

Not bad for a spiky saltmarsh weed that takes so many years to make up its mind about growing.

Recipes: *Grilled Asparagus with Egg and Bacon*

    2 lb. asparagus, cut into 6-inch lengths
    1 hard-boiled egg, chopped
    2 slices of bacon, cooked crisp and diced
    1 tablespoon olive oil
    1 teaspoon lemon juice
    1 piece of Parmesan cheese; salt and pepper.

Preheat barbecue grill. Brush asparagus with olive oil and season with salt and pepper. Place asparagus spears on to hot, oiled grill and cook 5–7 minutes or until crisply tender (skin will char lightly). Remove from grill on to a large serving platter.

Sprinkle chopped egg and chopped bacon over the top. Drizzle with olive oil and lemon juice. Season with additional pepper. Arrange shavings of Parmesan cheese on top of the asparagus. Serves four.

Whether boiling or steaming, it is a good idea to tie the asparagus in bundles of 10–12 spears for cooking, so they can be quickly removed from the water all at once. To boil: submerge in a large pan of boiling water, cover and cook for 3–6 minutes. To steam: stand the asparagus in 3 inches of boiling water, cover and cook for 3–6 minutes.

Roasting and chargrilling produce an intense smoky flavour quite distinct from boiling or steaming. To roast: pre-heat the oven (220°C), place asparagus on baking sheet and drizzle over a little olive oil. Roast for approximately 10 minutes, turning a few times depending on the thickness of the spears. Stems should be soft, not limp, tips should be slightly crunchy. To chargrill: toss the spears in a little olive oil and place in a ridged grill pan over a high heat. Grill the spears for 3–6 minutes turning once or twice.

Beekeeping: Home-made honey, great on your scones

In intervals between the downpours and hail showers of high summer, it has been noticeable that there aren't so many bees visiting our flowers. Bumblebees, yes, but honeybees, no.

Even on the white clover in the paddocks and the oilseed rape in neighbouring fields, it seemed that the bumblebees had the nectar and pollen entirely to themselves.

In front of the house, on the Erysimum Bowles Mauve which pluckily covers itself in wallflower bloom from Easter to Guy Fawkes' Day, there were a few honeybees, though half a dozen types of bumblebee were much more noticeable for much of the year. In late summer, they shared the flowers with an abundance of Red Admiral butterflies, plus a few Small Tortoiseshells, Painted Ladies and one or two Peacocks – a brilliantly colourful bunch of visitors who don't seem to realise that we are as far north as the Eskimo villages on Hudson Bay.

Maybe it is just one of those things, like memories of those long, hot summer days before global warming, but there do seem to be fewer bees about.

We have a professional bee-keeper only a mile or so away and slightly higher up the hill than our 300ft. His honey is popular, with stories told of minor Royals wandering around the back roads trying to track down this source of wonderful flavour. His bees do not seem to reach us, however, although another bee expert tells me that they should range up to two or three miles.

Thinking about the absent bees brought back boyhood memories of helping my Aunt Millicent to extract honey from the combs she retrieved, amid clouds of smoke, from her hives among the broom bushes – a tedious job it was, cranking the handle of a very manual separator. The main reason she kept bees was the wartime sugar ration granted to beekeepers. It all went into our tea and I don't think the bees ever saw any of it. They had to survive the winter on the honey she left in the hives.

Honey in the comb, rich and waxy, is still a popular addition to a home-made scone in our kitchen, so the idea of setting up a few hives seemed interesting. Particularly so since the clover mix of our sheep paddocks is definitely under-exploited by bees and we are surrounded by sycamore trees, rich in nectar.

What would it cost? Research produced some unpromising figures. Like most enterprises on the hobby farm, beekeeping is non-profit. According to figures from the agricultural diversification people, a basic kit of hive, veil, gloves, smoker and some bits and pieces, plus a six-frame colony of bees, would cost nearly £250. Extraction equipment might add £200, if I wanted to do that bit myself.

Returns, based on 50lb. of honey from my hive, would be about £75 a year, selling the honey wholesale.

The beekeeper I talked to – 'ten hives, great fun, five or six hours a week in summer, better than any other small livestock you can think of . . . ' – said I could sell it retail at £2.50 per lb. A clamp-down on honey from China (somebody found traces of antibiotics that should not have been there) recently cut off about half of our annual imports, he said, and that has boosted enthusiasm for the pure local product.

Another influence has been interest from alternative medicine enthusiasts, who have discovered the nutritional and wound-healing virtues of honey. They tell the story of the 20-year-old wound, infected with antibiotic-resistant bacteria, which stubbornly persisted in the armpit of a woman patient, so serious that she could not work and was in constant pain. An alternative therapist persuaded her to apply honey dressings to her arm. Within a month the wound healed and she was back at work.

The phrase 'some honeys' creeps in at this point, and researchers have suggested that particular sources, such as the tea tree (Leptospermum) give good medicinal honeys. We don't have any tea trees and our prospective honey might be better on toast than in your armpit.

There is also some demand for beeswax, the secondary product of the hive, and that would certainly fit in with the history of our place. For centuries, an annual payment of 7lb. of candle wax was paid to the local abbey in exchange for the right to maintain a chapel. It would be charming to revive the custom, using our own beeswax, even though the chapel has collapsed.

The great stumbling blocks, however, are mites and stings. The mites – the dreaded varroa – are still spreading among Britain's bees, killing off whole colonies. And the bee sting is not much fun if delivered in quantity.

It is said that less than one per cent of the population is genuinely allergic to bee stings, which is defined as the swelling extending more than two joints above the sting. Even the other 99 per cent, however, can feel distinctly groggy if stung by a few bees and it is not much consolation to know that 'regular honey bees will chase you no farther than half the length of a football field'.

Official advice from beekeeping organisations includes such phrases as: 'Make sure your bees are not of a strain prone to attack passers-by on sight from a distance.'

We are still debating the idea of a few hives, but over at the abbey they said a box of mail-order candles would be just fine.

Christmas Trees:
Don't fall for a fake fir

Christmas trees are a fashion business. About 40 million young conifers are growing in British plantations, all aimed at the Yuletide market, but many are doomed to be turned into wood chips because they are not going to be the 'now' tree when their harvest date comes around.

The fashion was first set in the 1840s when Queen Victoria and Prince Albert popularised the Central European Christmas tree, which came to be accepted as the Norway spruce. This was the unchallenged traditional tree for more than a century, but re-usable plastic trees moved into the market in the 1950s and 1960s, claiming to be more environmentally friendly.

As the supposed virtues of these loo-brush trees were disproved, with growers using the slogan Don't Fall for a Fake Fir, the Norway spruce made a comeback, with nearly 7 million being sold each December. In recent years other conifers have been promoted as the 'best' tree, most successful being the Noble fir and the Nordmann or Caucasian fir, both strongly marketed by Britain's 400 professional growers.

Our fashions usually follow American or European leads. However, with the fresh Christmas tree market, which is of such interest to smallholders, this has not happened because tree species which grow readily in those areas do not always succeed under British conditions. The Fraser fir is an example. It is considered to be superb in the north-eastern United States for such qualities as needle retention, fragrance and needle softness, but has not yet found a foothold here.

Because of the ten-year time lag in Christmas tree production, buyers can expect to see lots of Nordmann firs and Noble firs in the marketplace over the next few years. They have the advantages of a fine natural shape, strong branches, and needles which stay firmly on the tree. They also have a mild and clean aroma of pine disinfectant.

These trees can grow to enormous size in their native mountains – the Noble fir up to 250ft in Oregon – but their main enemies are tiny aphids, which can suck the life out of them when the young firs are only 10–15ft tall, so they do need to be grown with some care, given early shelter by nurse trees which are later removed to give the growing firs plenty of light and air.

They are quite fussy about climate and soil, being particularly suited to the glens of the eastern Highlands of Scotland. However, throughout the British Isles they are being planted by the thousand as the Christmas tree of the 21st century.

One area in which America has definitely set the trend is in shearing the growing trees to make them more dense and symmetrical, a development which does not suit some of the big firs, which sell on their natural shape. Each year a higher proportion of Christmas trees marketed in Britain have been clipped during growth to make them prettier. This is labour-intensive, but it pays.

The fashion trend is turning towards conifer species

which respond well to shearing. They should also retain their non-spiky needles and smell sweet. And an ability to carry a heavy burden of flashing lights and decorations without collapsing is another selling point. These attributes bring three of our common conifers into play – the Douglas fir, the Scots pine and the Lodgepole pine.

The Kew outpost at Wakehurst Place in Sussex has been conducting a trial of 17 conifer species over a number of years to establish their suitability as Christmas trees. Forty per cent of the planted crop was Norway spruce, the traditional tree, and 20 per cent was Nordmann (our own favourite) and Noble firs, seen as the trees of the future. But when customers were allowed to walk around the plantation and select their own, the clear favourite was the Douglas fir.

This North American fir has been grown commercially in vast numbers in this country and its requirements are clearly understood. As a Christmas tree it has the virtues of highly-scented foliage which does not drop, and it grows into a dense shape when pruned.

Scots pine and Lodgepole pine do not look in the least bit like the Norway spruce of childhood memory. They do, however, hold their needles even if kept dry in the house for weeks, and they form a remarkably bushy and symmetrical tree if skilfully trimmed in June.

The shift towards widely-available plantation conifers whose husbandry is straightforward, plus the emphasis on trimming, makes the Christmas tree market quite promising for the hobby farmer. The main requirement is low-cost summer labour for weed control and trimming, and this is often available within the family enterprise.

Another perceived trend which gives an advantage to the small-scale grower is the desire of buyers to select a tree at the farm gate, perhaps even to cut or dig their own from the plantation, with all the members of the purchasing family helping in the selection. A variety of species and tree ages make this a more interesting process, and pick-your-own is definitely the way to have the customers coming back again, year after year.

If the hobby farmer's available site is rich English farmland, the battle against weeds is going to be a key problem. Even under good conditions, large numbers of young conifers are smothered or die, with only 60-80 per cent making it to saleable size in many commercial operations. So allow plenty of space to get in among those growing trees with mower and sprayer – spacing of 7ft x 7ft is often recommended as a minimum – and be prepared to use weed-killer and pesticide. Organic growing of Christmas trees is hard work. And be sure to plant in straight lines to make the mechanical operations feasible.

Shearing/trimming involves considerable skill, starting with some tidying up of double leaders and stray growths in the second and third year, becoming an annual process from then onwards until harvest in years 8–10 and upwards. The ideal Scots pine, as trimmed, is about two-thirds as wide as it is high. A tree 6ft high and 4ft wide at the base would have this ideal taper of 67 per cent.

The timing of shearing of the pine species is quite critical, taking place immediately after leader growth stops and buds are set in June. Firs and spruces can be sheared at any time after height growth ceases and onward into the winter. Wakehurst's spruces are even disbudded by hand. One of the reasons for the popularity of the Fraser fir in North America is that the species can be clipped like a hedge and it has an abundance of buds ready to break into new bushy growth.

The equipment used for shearing is not exactly family friendly. Most American growers use machetes and knives, plus Kevlar gloves and strong leg guards so that a misjudged swing does not take off a kneecap. The practice in this country seems to favour shears. Whatever the tool, lots of paraffin is needed to wipe the blade and prevent build-up of resin.

On the subject of trimming, it is worth noticing that some of the most neatly sheared trees I have seen are standing around the unfenced edges of commercial plantations of conifers where they have been browsed by deer – for example, near the London-Glasgow motorway at Beattock Summit in the Lowther Hills, just north of Moffat. Deer are not always so neat in their browsing habits, however, and have to be kept at bay with fences. The same applies to rabbits. Keeping these pests out of the plantation can be a major cost.

Mixing domestic livestock with growing Christmas trees is rarely successful. A flying flock of sheep might graze their way along the mowed grassy strips for a few weeks, but once they start pushing their way into the trees and breaking branches they outstay their welcome.

Mixed plantations are a possibility, for example where it is planned to establish an orchard, a crop of Christmas trees can be grown in the 30ft gaps between the young apple trees.

The final stage is selling the tree. As the Americans say, you've got to sell Christmas, not just a tree. Marketing effort is rewarded. It takes more than the scrawled sign: XMAS TREE'S. One of my neighbours had great success last year with a single-price policy. All his trees, from 3ft to 30ft were priced at £10, pick and haul your own.

Cider Apples: The end product is innocent but powerful

Apples from the old trees in the orchard used to be one of the most valuable products of any country house garden. But these days the crop is likely to fall and rot in many English orchards because these days nobody is quite sure what to do with them. They are too tart to eat, too small to peel and cook, too unpredictable as livestock food.

The answer, of course, is to use the crop for its original purpose – to make cider. It's a slow process creating your own apple brew, but until the long-promised summer arrives when we all have abundant grapes hanging from our garden vines, it is one of the most fascinating ways of providing home-grown refreshment.

Cider has had its ups and downs as an English drink. At some periods, it has been highly esteemed, with West Country ciders being sold in London at prices which were exceeded by only the finest of French wines. At other times, it has been derided as the roughest of rural rotgut.

Summer is the time to sample some traditional ciders, discover the types you like and work out how to make something similar from your own apples. As with beer and wine, there's a huge range of ciders, from bitter-flat-and-cloudy to clear-and-bubbly, all with a long history behind them.

Home brewers can make real ale at any time of year, but cider is seasonal; you make it when you have ripe apples, falling off the trees, from September through November. And unlike beer, the final product draws all its character from your home-grown fruit. This varies from year to year, so cider is often call the Wine of the West, with recognisable vintages.

There are special varieties of apple which are grown for cider, most of them with wonderful names such as Foxwhelp and Kingston Black, and many are confined to just a few parishes, some now very rare. Bitter-sweet or bitter-sharp, these cider apples can be blended with sweet dessert varieties to create the desired balance in a complex juggling act involving all their subtle tastes and aromas.

The apples go through two major processes: crushing and pressing. The crushing is heavy work if you are dealing with large quantities of fruit and the apples do have to be crushed. Even industrial machinery finds it almost impossible to get the juice out of apples unless they have previously been squashed and knocked about fairly thoroughly. One home cider-making handbook recommends a steel bath and a length of four-by-four timber as a suitable crusher for the enthusiastic (and muscular) amateur.

The crushing is followed by pressing, squeezing the juice out with a piece of machinery which resembles a lorry jack. This juice is then put into a barrel and allowed to ferment – using either the yeasts present on the skins of the apples or adding some champagne yeast to speed the process. There are lots of ways of tinkering with the basic timetable, which involves two separate fermentations in autumn and spring, but if you maintain the ancient schedules you will be drinking your own cider 'when the first cuckoo calls'.

The creation of an old-fashioned country cider was described in detail by Mrs. Gennery-Taylor, writing more than 40 years ago: 'For the best cider, a mixture of different varieties of apples is best. Those usually chosen are non-keepers, small sour or windfall, with, if desired, a few crab-apples. An odd rotten apple in a large number is permissible, but otherwise they should be sound.'

She went on to explain that the juice should be put into a wooden cask – a 30 gallon ex-brandy cask is ideal for first-class good keeping cider. Base your calculations on the fact that a ton of apples makes about 150 gallons of cider, therefore a hundredweight makes approximately 7.5 gallons. After about 48 hours the apple juice will start to ferment and white froth will bubble up through the bung hole. This will continue for about three weeks. Never bung up the hole while fermentation is going on

When this fermentation has almost stopped, add sugar. Using sterile plastic pipe, siphon out sufficient juice to dissolve the required amount of sugar – 2–4lb of sugar per gallon in the cask, depending on how sweet you want the cider. When quite dissolved in warmed juice, allow to cool, then return to the cask. (There is some difference of opinion about heating the juice: if you get it wrong, this can result in a cider like cloudy apple soup.)

Owing to the addition of the sugar all the sweetened juice will not go back in at once. During fermentation, which will go on for about two weeks, the quantity of liquid in the cask reduces so that you can add the surplus gradually (as space permits). When fermentation has nearly finished, if all the 'juice and sugar mixture' is not in the cask, siphon out enough juice to allow this to go in. Bottle what you take out and use to keep the cask full while the cider is maturing – the quantity reduces during this process. Developing airspace in the cask will otherwise allow bacilli to breed and turn the cider acid.

When the juice has completely ceased to 'bubble up', bung the cask tightly with either cork or wood, and leave for eight months.

Mrs. Gennery-Taylor added: 'Innocent to taste but powerful – up to 15 per cent alcohol can be achieved.'

There is much confusion in the world of cider. For example, 'scrumpy' can be either the most basic of rough ciders, made from windfall apples, or it can be a top-of-the-range brew. And 'apple wine' is undefined: it lacks the bite – bitterness plus sharpness – of most ciders but usually has higher alcohol content and uses wine yeasts.

The tendency was for English connoisseurs to prefer bitter West Country ciders, although there has been an increasing market for the French style, light and bubbly. Cider historians have recently discovered that the original 'champage style' drink was an English cider, created when improved methods of glass-making, using coal instead of charcoal, allowed higher pressure to be captured in the more trustworthy bottles.

Famous apple varieties for cider-making have included Foxwhelp in Herefordshire, Sweet Alford and Woodbine in Devon, Morgan's Sweet in Somerset and Kingston Black. The National Fruit and Cider Institute ran extensive trials in the mid 1930s leading to the widespread use of Yarlington Mill, a seedling raised in Somerset at the end of the 19th century.

The big cider factories now buy fruit from France and also import concentrated apple juice from abroad. However, the best ciders are still produced by small farms using their own cider apples. The Campaign for Real Ale (Camra) now does a great deal to promote good producers and October is their Cider Month.

With so many variables in the ingredients and production of cider, there is no standard recipe which produces perfect results. To learn a little bit about the subleties involved, you can start with juice and yeasts from a home-brew supplier. Even using these, the process will take about six weeks from first fermentation to first glass.

The high-speed recipes usually knock out the wild yeasts, using campden tablets (sulphur dioxide) as in home wine-making, and move things along at various stages by treating the fermenting juice with finings to remove cloudiness. A more traditional cider uses ale yeast, while Normandy ciders use hock or champagne yeasts. These can be re-fermented in the bottle like sparkling wine – and don't let anyone tell you they are not traditional in England, because they have been around for more than three centuries.

But the most fascinating variable is the apple itself. As the Camra results show, there is still something about western apples which appeals to the taste buds of connoisseurs.

If you don't have those old apple trees at the foot of the garden, you might like to try planting a few cider varieties and making plans for the next 10 years and beyond. Several specialist nurseries sell young apple trees in pots, suitable for planting at any season.

Cobnuts: Tradition which fits into the hobby farm

A grove of cobnuts is one of the most charming ways of turning a field into a productive orchard without too much labour. And a 'plat' of nut trees has the added bonus that income from pick-your-own harvesters could repay much of the cost.

The growing of the hazel nut, cobnut or filbert was once widely practised in the south-east of England but gradually withered away until the 7000 acres in cultivation in Edwardian times had shrunk to less than 250 acres. But several new orchards are now being planted and there's burgeoning interest in this fascinating traditional crop which can thrive alongside other hobby farm enterprises such as free-range poultry.

A fistful of cobnuts, a crisp pippin apple, a piece of Stilton cheese and a sip of port was long regarded as the ideal snack for an autumn evening. If you are growing your own apples and nuts, the investment in port and cheese is really worthwhile!

Starting point for your nut grove is the young plant from a nursery – priced from about £3 upwards and standing perhaps 2ft tall at two years old. These young hazel bushes are disappointing at first sight. They are puny things compared with apples or soft fruit. However, they get going in year three and by the time they are 10 years old they should each be producing nearly 10lb of nuts annually.

The bushes are put in at spacings of approximately four yards each way, slightly less if you are attempting to make a profit, but leaving lots of space if you want to grow grass underneath. The plat is kept tidy by whizzing around with a ride-on mower. Poultry are fine as an under-crop but hazels are badly knocked about by sheep or ponies.

Cobnuts are quite hardy and they even benefit from a breezy site because the early spring pollen from the catkins is spread around by the wind. The hazel/cob/filbert – several species and hybrids of *Corylus* or one super-species, depending on your botanical source – is not strongly self-fertile and it is advisable to plant a mix of hybrids to reduce the proportion of empty, unfertilised nuts. The only way to detect these (although squirrels do it unerringly by eye) is to put the nuts in a bucket of water – the empty ones float.

Young cobnut bushes grow from a short trunk from which half a dozen branches radiate at about 18 inches above the ground. The trunk does not live for ever, but the small tree keeps renewing itself with growths from the base, so that 100-year-old clumps are still healthy and productive.

Pruning can be elaborate, done by hand, or swift and brutal, done with a tractor-mounted cutter blade. Each style has its devotees. Like pruning roses, the style depends more on the pruner's personality than the final results. Almost anything works. The aim is to keep the small trees down to about 10ft in height, so that you can pick the first nuts at the end of the summer without the need of ladders, although later in the season they are shaken out of the bush.

The pruning season runs from mid-winter to March and hazels are trimmed every year. March pruning is often recommended because disturbing the trees when the pollen begins to fly is helpful.

Traditionally, there was also a summer pruning called 'brutting', during which the minor side growths which carry female flowers in spring are broken about halfway along their length, using the back of a knife blade, and left to hang. This was believed to reduce the amount of secondary growth in the bush, but labour costs have done away with that stage of the process.

Long wands growing from the base of the hazel can be used for lightweight poles, replacing bamboo canes around the garden. The timber of mature hazel trees is quite strong but there is never very much of a trunk, so the heavy, twisted wood has few traditional uses apart from umbrella handles. The fast-growing, pliable rods which spring from the bottom of the trunk are much more useful and have had scores of vital functions over the centuries, from basket making to fish traps and bird cages.

True aficionados of the hazel nut start eating them off the trees in late August or early September, when they are still unripe and milky. The filbert variety gets its name from St. Philbert, whose day falls on August 22 at the start of the harvest.

The flavour of the nuts is very light and delicate at first and does not intensify until a month later. Some people like them best when the enfolding bracts – longer in the filbert varieties than in the chunky cobnut – are golden green in colour, while others would prefer to wait until October when the crop is brown, with the kernels firm like the familiar hazelnuts in a bar of chocolate.

Cobnuts are worth sampling as a 'fresh fruit' rather than

a dried nut. To try them year round, contact:

> Allens Farm Kent Cobnuts,
> Allens Lane,
> Plaxtol, near Sevenoaks,
> Kent TN15 0QZ
> (www.cobnuts.co.uk)
> 01732 812 215,

who supply mail order from cold storage and also have outlets at Covent Garden Piazza, Borough Market and Fortnum & Mason.

The most widely cultived variety of hazelnut in the southeast is the Kentish Cob which is actually a long-husked filbert with a large pillow-shaped nut. Because of the need for cross-fertilisation with other varieties, bushes grown from its seed (from the nuts, that is) are not pure Kentish Cobs and replacement orchard bushes have to be propagated by layering, tying down two-year-old branches and encouraging them to develop their own roots.

Many nuts from Kentish Cobs are, however, used to produce nursery stock for replanting schemes and the resulting mongrel trees, with genes from across Europe and the Near East, are passed off as 'native' hazels in various worthy projects throughout England.

There is no firm rule to distinguish where the cobnut ends and the filbert begins, although one botanist suggests wryly that 'a filbert is a hazel with a college education". Whatever you call them, there are numerous varieties in nurserymen's lists. Some of these *Corylus* hybrids are quite ornamental, particularly the purple or red clones of the filbert type. Hillier's Manual, one of the classic reference books on trees and shrubs, says the purple form 'rivals the purple beech in the intensity of its colouring.' The nuts are good too.

Cobnuts are often marketed fresh, not dried like most other nuts such as walnuts and almonds, and there is a growing trade in pick-your-own in summer. They appear in the supermarkets when in season, from the middle of August through October, although stored nuts may be kept in a cool room or the fridge until Christmas or beyond. Because of their keeping qualities, they were often an important part of mariners' stores.

After all the good news about hazel nuts, there is a foot-note. Grey squirrels, which were unknown in this country when early Victorians planted the Kentish plats, are a serious pest of the crop. If you have squirrels, you will have to be prepared to control them if you want a worthwhile harvest.

Varieties

Kentish Cob: A reliable crop, in clusters of 2–5 nuts, relatively hardy, with excellent flavour; also available as Longue d'Espagne; pollinated by Gunslebert, Cosford and Merveille de Bollwiller.

Gunslebert: hardy, vigorous and very productive producing medium-sized nuts: pollinated by Cosford and Kentish Cob.

Merveille de Bollwiller: Also called Hall's Giant; a hardy, vigorous and productive variety with large nuts; pollinated by Cosford, Butler, Ennis and Kentish Cob.

Butler: A large mid to late season nut, hardy, vigorous and a heavy crop; a short-husked variety that de-husks freely when ripe.

Webb's Prolific: Large nuts to eat raw with cheese and port. Sold by Marshalls (see appendix), who also sell the Red Filbert.

Cosford: Tall, upright tree with thin-shelled sweet nuts. Grown as pollinator.

Crayfish: Confusing rules for the lobster lookalike

Just over 20 years ago, the Government was giving grants to any smallholder who wanted to start a crayfish farm, and officialdom was enthusiastic about the importation of signal crayfish from North America.

Today the official view has changed: the signal crayfish is an evil import on a par with the ruddy duck and giant hogweed, and every effort should be made to destroy it.

For those of you who have not been paying attention, this somersault of opinion has all happened very quickly. For the hobby farmer trying to make a pound or two from his half-dozen acres, it is all very confusing – and very disappointing, because there is a real demand from the restaurant trade for pond-grown crayfish, and they can show a profit.

The crayfish which has caused all the bother is the signal crayfish, an American relative of our native freshwater species, the white-clawed crayfish, whose stronghold was in southern chalk streams. The native was finding life difficult – allegedly because of pollution, disturbance and other man-made problems – but then the signal crayfish escaped from the early crayfish farms and proved very rapidly that it could survive in our welcoming streams and ponds.

It not only survived, it thrived exceedingly. Within a few years it was colonising whole river systems across the South of England and making overland jumps into ponds and reservoirs. Soon this lobster lookalike was being dredged up in amazing quantities. Typically, a setting of a couple of dozen traps overnight might catch half a hundredweight, and a week later the same again, all summer long.

The men from the Department were (and are) very upset. The signal has ousted the protected white-claw and has also introduced a fungus disease, a sort of crayfish plague, to which it is largely immune but which wipes out whole colonies of the native.

The signal crayfish is a success, and it seems that environmentalists would rather fight losing battles. They are wholeheartedly on the side of the native, which is now hiding behind some serious legislation.

The result of the complete about-turn in officialdom's view of the signal crayfish is that farmers who were given money to go into the business of breeding them, now find that it is a criminal offence for their would-be customers to buy crayfish for stocking ponds across most of the country. The only outlet is direct to restaurants.

And restaurants love them. The crayfish is a grey-green, freshwater mini-lobster, bright red when cooked, with a body small enough to fit on the palm of your hand. It is very tasty, and although they have never made much of an impact on the British menu – I remember a pub at Hungerford which treated the native form as a curiosity rather than a delicacy – crayfish are highly valued elsewhere, particularly in Scandinavia and North America, where they are often known as crawdads.

The world record signal crayfish was taken from a

Nottinghamshire lake and it weighed less than half a pound, so they are more like scampi than lobsters on the plate, and the flavour can be like either, depending on the cooking technique.

The Swedes devote whole summer festivals to crayfish. The mounds of bright red cooked shellfish, often simmered in beer, are flavoured with dill and served with salad and dips. The French add mayonnaise. British chefs go for a moules marinieres approach, serving them hot, with cream and shallots.

The Americans enjoy a boil-up recipe, which is based on a gallon of water. Add enough crayfish to the briskly bubbling pot so that they are just covered by the water, then pour in half a cup of salt and half a cup of sugar. Boil them for 10 minutes, followed by half an hour's cooling time in the same water. Shell and serve with garlic butter.

A crayfish farmer near Dorchester told me: 'London restaurants just can't get enough of them. We're sending off 20 kilos every week, from May to September, at £20 a kilo.

'People love the idea of eating a freshwater shellfish which is totally organic. The crayfish are not fed on any-thing artificial, they feed completely on natural vegetation that's already in the pond. We do not give any supplementary feed at all – that was one of the reasons we put them in our ponds, to browse the weeds.'

She says that crayfish got rather a bad name when they were originally marketed in this country because fish farmers did not take enough trouble over grading them: 'Some people thought you could just take a fyke-full of crayfish out of the net and throw them all into the pot. We grade them, making sure that we send out only select crayfish that have just moulted their shells, all perfectly clean, the right size and not carrying eggs.

'Also purging them is very important. We always purge them for two or three days, feeding them on potatoes, so that their intestinal tract is completely cleaned out.'

She is just sorry that the present law makes it so difficult for other people to get into the business, because there seems to be a big enough market for everyone. She told me: 'We were advised and encouraged 16 years ago to start with crayfish, using ponds on an area of waste land on gravel. A river runs through our farm, so if we dig a hole it fills with water, perfect for crayfish. It's unfortunate that now we just cannot sell them to people who want to breed them, but we have to say no.'

The rules about fishing for wild signal crayfish and conserving the native species have been confusing in recent seasons, with DEFRA trying to construct a stable door and bolt it decades after the damage had been done. This was illustrated by the farm diversification website of the Scottish Agricultural College, which devoted three pages to crayfish farming, including details of the financial assistance available, and then noted it is 'unlikely to be approved by the appropriate authorities'.

Meanwhile, the effect that the signal crayfish is having on our rivers and ponds is disputed. Some say this horror crustacean is muddying the waters of clear streams, under-mining the banks, chasing away water voles and destroying everything from caddis larvae to fish eggs. For others, the only effect (quite unexpected) is that signal-colonised waters seem to produce bigger wild trout.

A great swathe of England, roughly the area south of the line from the Wash to the Severn, has become the kingdom of the American crayfish, and no one is seriously suggesting that the species can be eradicated. North of that line, the white-clawed is hanging on. Permission might be granted for a well-fenced crayfish farm in the South, but sources of stock are problematical under present legislation.

So is there a farmed future for our American crayfish?

The other day I was offered a sandwich in a City bar – fresh crayfish and avocado on wholewheat, £5.20. At that sort of price, this much-reviled form of pond life is going to be farmed somewhere.

But if the environmental purists have their way, it is going to be somewhere else.

Donkeys and Mules:
Familiar pets are great workers

Not far from our ten acres, there is a man who keeps everything. James not only keeps everything, but he keeps it on four acres. One horse, two ponies, a donkey, llamas, sheep, a goat and geese, plus stray bantams...all apparently happily grazing together.

How is it done? What skills do you need to mix and match all these forms of livestock in one field without open warfare breaking out?

Our own experience suggests it is not as easy as it appears. Our mixing of sheep with horses ended in a particular tragedy. A colt who was a bit too free with his heels struck a Castlemilk Moorit tup, who was being aggressive in his very formal, head-to-head way, and left the poor ram paralysed, with a broken spine.

'It happens all the time,' said the vet. 'Sheep and horses are not very evenly matched, you see.'

I asked our neighbour about the secrets of keeping different kinds of stock together, certainly not evenly matched, without having dramas that involve calling the vet. James thought about it for a moment and said: 'You need good hay, regular doses of wormer, big sheds and a eye for trouble. And the donkey helps.'

Before asking for details of the sheds, hay and medication, I wanted to know what the donkey had to do with it.

'He's an old gelding – well over 30 now – and he's a calming influence on all the others. He's hardly done a day's honest work in all those years, but he does seem to be a peacemaker. Every one of those animals in that field thinks he is their special friend. The sheep like him particularly – he chases off stray dogs.'

The donkey in the British landscape is not often seen doing an honest day's work, although in the poorer and drier parts of the world he is the most overworked of labourers, carrying huge loads up the mountains and across the dry fields of Peru, Egypt or China. In this country, he is usually living in quiet retirement. There are still donkeys around which once worked in Irish peat fields and even a few with memories of the coalfields. They can live to 50 or 60 years of age, seeing off several generations of their companion horses, and sometimes their owners too.

I asked James how he came to own a donkey. 'I inherited it,' he said. 'The beast came with the field. He was thought to be quite old then, by horse standards, and that was nearly 20 years ago. He's been eating my hay ever since.'

The donkey's virtues include the ability to live frugally on a simple diet. In fact, more donkey health problems are caused by overfeeding than almost anything else – they must not be allowed to get fat. And they are the best of companion animals for a wide range of hoofed stock, seeming to have a particular talent for calming nervous horses.

The animal's drawbacks range from being famously stubborn to having that hee-haw voice which rivals the peacock as a cause of social embarrassment. The screech-honk, screech-honk might be part of the atmosphere in a Greek village but in rural Britain waking everyone within a mile is not considered amusing.

People are, however, very sympathetic towards the donkey itself, even if they might fall out with the animal's owner because of the noise. The humble ass is often seen as rather downtrodden, sad and exploited. Elderly ladies who have hardly ever seen a donkey frequently leave their entire fortunes to the Donkey Sanctuary in Sidmouth, Devon. It is one of the richest charities, with nearly £15 million being donated every year to its good cause, looking after 3200 donkeys and promoting better management of these animals around the world. For many years the Sanctuary has appeared in the top 100 charities, rated by value of donations.

Dr. Elisabeth Svendsen, who founded the Sidmouth farm, has also compiled a useful book on caring for donkeys (The Professional Handbook of the Donkey, see appendix). She makes the point that the horse has enjoyed the benefits of a great deal of research into its diseases and management, all financed by rich owners. The asses' poverty-stricken handlers in Africa, Asia or South America do not know that their beasts are being crippled by starting heavy work too young and and are burdened by internal parasites which will kill most of them before they are a dozen years old, even though they have the potential to live for another 20 years or more.

Britain's pets, by contrast, live very well. And they are having something of a resurgence at the moment, lining up in the show ring at scores of events around the country. The British Donkey Breed Society has more than 1000 members and celebrities seem to love the long-eared pet.

Donkey breeds are mainly defined by size. From the islands of Sardinia and Sicily come the Miniature Mediterraneans, which measure under 36 inches at the shoulder.

They have their own breed club, with pedigrees going back many generations to animals imported into the United States in 1929 by Robert Green, a New York stockbroker who established a donkey stud farm in New Jersey. His enthusiasm for his little animals was summed up in his description: 'Miniature donkeys possess the affectionate nature of a Newfoundland, the resignation of a cow, the durability of a mule, the courage of a tiger, and an intellectual capability only slightly inferior to Man.' Now that's real enthusiasm.

Although they are no longer common in the Mediterranean islands, there are thousands of miniatures in North America and several hundred have been imported into this country, establishing the breed as one of the most distinctive in the donkey family. They are undoubtedly cute and come in all the grey, brown and spotted variations of other breeds.

At the other end of the scale, there are the big ones, such as the extremely rare French Poitou donkey, raw-boned and tangle-haired, which stands 14-plus hands. Numbers of this variety had dropped to less than 100 a few years ago, but it is having something of a revival and there are now about 400 of the distinctive heavy-headed, shaggy black giants.

The big American breed is the rather stylish Mammoth Jackstock, also standing at 14 hands or over, much used for mule breeding. In between are the Standard and Large Standard.

Whichever type of donkey you find interesting, there are some distinctive features which have to be kept in mind when making plans to keep one on your hobby farm. Donkeys need a friend, preferably another donkey, and they need shelter. The species is Africa's only major contribution to the list of domestic animals and its origins in Nubia mean that it is basically ill-adapted to wet weather. Its coat is thick enough to keep out the cold but it does not have natural oils to repel water – it soaks up rain like blotting paper and takes a very long time to dry out. Donkeys really do not like water, which you will discover if your children try to ride one through a stream, and one of their luxuries is a roll in a dust bath.

Donkeys' hooves are hard-wearing, even though they look so peg-like compared with horses' wide hooves, and they are not usually shod.

Mares and geldings make good mounts for small children. They are patient and affectionate. British donkeys tend to be under 11 hands (44 inches or 112 cm) and can carry up to 8 stone (112lb or 50kg). Because they do not do high speeds and dislike jumping, their usefulness is limited to the youngest beginners, although you could invest in a trap and go on to try some driving.

James told me that he finds the donkey interesting but basically not much use around the place. However, he really fancied having a mule, if he could find enough space for it. 'That's the best working animal ever invented,' he said.

I've covered a few miles on sure-footed mules, scrambling up and down some of the world's steeper paths, from the Ecuadorian Andes to the Italian Alps, and I would agree. A good mule (illustrated above) is a very special beast, and for many jobs it is better than either of its parents, the donkey stallion and the female horse. Curiously the reverse cross, the hinny, out of a donkey mare, is not nearly so strong or so talented.

The mule has been described as a horse with a donkey's head and feet. However, there is more to it than that, and although almost every feature of the mule is described as 'weaker' or 'narrower' than the corresponding part of the horse, whether it's the rump, quarters or sides, the complete animal is renowned for strength and stamina, with a much longer working life than the horse. Through thick and thin, the mule obstinately soldiers on, rarely going sick or lame.

James recently saw some mules hauling logs in Mexico. He thought they were inspiring: 'I'd get rid of almost every other animal on the place if I could find a good mule. They're impressive.'

My neighbouring smallholders mostly dream of buying a JCB or some other piece of high-tech equipment and hiring themselves out as contractors. James now has a brighter vision of a future as a muleteer, bringing logs skidding down through the forest or carrying equipment for shooting parties. I am watching his four acres with interest.

Ducks for Eggs: Peasants love the birds that lay the money

The first thing to decide when you consider having a flock of laying ducks is: Do I *really* like eggs?

When you have ducks you are going to have an awful lot of duck eggs. And those ducks lay their eggs with a casual enthusiasm far removed from the serious way in which laying hens go about the process. They drop them on the grass, in the feeding trough, in the duck house, in the pond and sometimes even in the nest boxes. You are going to feel as if you are up to your knees in eggs.

Well-fed ducks of a specialist laying breed, such as Khaki Campbells, are quite capable of producing 300 eggs a year – some have even topped 350. A ten-bird flock on your pond could be presenting you with 3000 eggs every 12 months.

We have a smallholder neighbour who is rearing her children on a foundation diet of duck eggs and Tesco's white loaves (four for a pound and keep the change). They are all looking well on it, so far. There is no nutritious food which is easier to produce than a duck egg.

A much-travelled crofter in the far North-west once told me that the way to approach any small-farm problem was to ask: What would a Chinese peasant do in this situation? If a billion of them are asking the questions, somebody must be coming up with some answers. And the Chinese say: If in doubt, keep ducks.

They love ducks because they make money.

Throughout South-east Asia, which is the original home of our laying breeds, the traditional method of keeping the birds is on a free range, herded flock system. The ducks are penned overnight but spend the day roaming around as a group, finding much of their own food in canals, rivers or the foreshore. If you've got access to a mile or two of saltmarsh and riverbank, and plenty of time on your hands, shepherding 500 ducks could be the way to go.

It's interesting to see that our Khaki Campbell has been proved to be the most profitable breed in Vietnam, after researchers compared the profitability of 6000 ducks on the Red River Delta. The Gloucestershire variety was adaptable enough to come top of the table, especially on free range on the delta marshes.

For most of us, though, the laying flock is confined to a couple of paddocks and a pond, with a fox-proof shed. And the shed is important. Where there are foxes, they will be the main threat to any flock of ducks, which need to be shut in somewhere safe at night, or at least given the protection of some water around their night-time base. A house on an island in the pond is almost secure enough, but cannot be 100 per cent guaranteed because a hungry fox will swim if tempted by the smell and sound of ducks and will certainly nip across the ice if the pond freezes.

Once those foxes and the egg-eating crows have been held at bay, the duck is a good investment. She is extremely hardy, easy to rear, unfussy about food, and remarkably disease free when compared with the laying hen. She is also a pleasant, friendly character, with an easy-going personality: people who keep ducks always comment on the fact that they are more fun to have around than any other kind of poultry.

The duck house does not need to be an elaborate affair: lack of ventilation is likely to be a bigger problem than exposure to the elements. Ducks like the elements. They roost on the floor, on straw, shavings, sawdust or whatever. They are mucky birds and cleaning out regularly is an essential chore. There is no need for a small pop-hole entrance – just use the shed door because all the ducks will emerge in one packed rush when you let them out.

The females lay in the morning, in theory, and if kept shut in until 10 o'clock or so, you should get most of the eggs in the house. In practice, they lay everywhere and you may find yourself racing the magpies and crows to pick up some of the eggs.

Ducks' tendency to turn any patch of lawn into a mud-bath is their only real drawback. Be prepared to give your birds either a concreted hardstanding area, hosed down frequently, or lots of grassy space. They do little damage in the vegetable and flower garden, apart from nipping off seedlings, and a great deal of good when they devour slugs, so you may be able to let them wander the gardens in winter, switching to the orchard or paddock in summer.

Their diet can be poultry layers' pellets in the morning and wheat in the afternoon, with appropriate starter and grower rations for young birds. They truly are easy to rear. If you buy 100 day-olds, the hatchery may give you two or three extra ducklings for luck and it is quite possible to rear more than 100 per cent of your original purchase.

In the old days, the cottagers of the Vale of Aylesbury used to rear their fat ducklings under hens in 'every room of the house, including bedrooms, a single room accommodating in some cases two or three hens with their families, separated by boards placed on edge. The noise at feeding times was deafening."

How much water is necessary for rearing these ducks is a much-discussed point. Ideally, they would have a running stream, or a pumped water supply in a series of troughs or ponds. However, the small laying breeds have long been kept with no more than a generous supply of drinking water, often a basin under a dripping tap, and they lay well and thrive under such conditions. They certainly do not need a pond to ensure fertility.

When Adele Campbell, of Uley, Gloucestershire, created the Khaki Campbell breed just over 100 years ago, using the Indian runner duck, mallard and the French Rouen variety, she did so in pens with drinking water only.

The Campbells are still the main laying breed today. Their productivity developed from the egg-laying powers of the Indian Runner, the familiar 'hock bottle' bird so often seen at country shows and fairs being herded by collies. The stylish and fast-moving Runner is still very popular – more than 20 per cent of the ducks at auctions or penned at shows are likely to be Runners, usually white, trout or fawn-and-white – but the breed's laying abilities have not been nurtured very well by breeders concentrating on the penguin shape. A 250-egg duck is said to be a good layer today and an embarrassment of eggs is not so likely if you buy show-pen stock.

There is a great deal of interest, however, in bringing the laying powers of light duck breeds back to the levels of the 1920s and several breeders are currently selecting for more productive strains.

'Of the various breeds of ducks, the greatest forager and most prolific layer is byond doubt the Indian Runner' – so wrote Lewis Wright in his monumental *Book of Poultry* in

Edwardian times. And the emphasis on foraging is of interest today, because the hobby farmer is looking for a laying bird that will range over park and paddock to find some of its own living and qualify for the genuine tag of 'free range'.

For those with water and grass, ducks are ornamental and productive. A drake and half a dozen ducks will give a steady supply of big, white eggs year-round, with new generations of ducklings coming into lay at about six months old, without any artificial light.

Small numbers of replacement stock can be readily reared under broody hens: Indian Runners and Khaki Campbells cannot be relied upon to incubate their own eggs. (When you think about it, 350 eggs a year does not leave much space in the calendar for rearing a brood!)

As far as choosing a breed is concerned, the true eggers are Campbells – available not only in the classic khaki colour but also in white and dark – and the Runners, which come in nine recognised colours and patterns of fawn, brown and white. If you are Welsh and feeling patriotic, you can work with two fine breeds, the Welsh Magpie and the Welsh Harlequin, both dual-purpose for meat and eggs, and both very handsome. The white-frosted Harlequin ('a Campbell which has run into a snowstorm') is one of the prettiest of utility ducks.

Others which may turn up at your local auction market include Buff Orpington (a duck breed as well as a large fowl), the attractive Abacot Ranger, Swedish, Saxony and Silver Appleyard. As a rule of thumb, the bigger the breed, the fewer the eggs.

Easter Egg Chickens: The story of the mysterious Araucana

Blue eggs are a red-hot item in the upmarket supermarkets of Britain. Just how hot I discovered the other day when an eager shopper elbowed my wife out of the way to grab the last half-dozen from the shelves of a Chelsea store and headed for the checkout with cries of delight as if she had just won the lottery.

One supermarket executive is quoted as saying that 'blue eggs are the best thing since canned custard,' and they are selling at premium prices, as we could see for ourselves in Waitrose in the Kings Road.

The fashionable blue egg from South America has come a long way since its first mention in *National Geographic* magazine more than 70 years ago. Its discovery is a wonderfully complicated story involving Indian tribes and Spanish conquistadores, cholesterol and fatal earrings, the Isle of Mull and a shipwreck, and a mysterious bird called the chachalaca.

Some of the tale is true, most of it isn't, and the chicken at the heart of it is keeping its secrets to itself.

The breed is the Araucana, a rather quiet and very hardy

bird, often seen as a bantam and sometimes as a large fowl, not quite as large as most modern egg-layers, usually with a tufty crest and often with curious feathered 'earrings' on the sides of its head. It comes in most chicken colours, one of the most popular being a blue-grey lavender shade. And it lays blue eggs – the pale, bright colour infusing the whole shell, inside as well as out, although the colour of the egg inside is not affected.

When the Araucana came to light in the 1920s, the story was told of this ancient breed which had been discovered among the Mapuche Indians of central Chile by Dr. Rueben Bustos, who had rescued some of the rare birds and bred them in his back yard. He reported that they had been known to the first Spanish missionairies who ventured down the Pacific coast in the 1500s. The conquistadores were said to be amazed that the Indians had such wonderful birds.

Since the wild ancestor of all domestic poultry, the red junglefowl, is found only in southern Asia, the presence in sixteenth-century South America of a unique domestic chicken would suggest either some extraordinary prehistoric trade links across the Pacific, a misunderstanding, or a hoax. Scientists are still arguing about that, although they quickly discounted the idea that the Araucana was created locally from a jungle gamebird, the chachalaca.

Archaeologists have not found any convincing evidence that domestic fowls were present in the Americas in pre-Columbian times. On the other hand, there is a great deal of written evidence that the exploring Spaniards found village chickens wherever they went, often very different from any with which they were familiar, and it seems extraordinary that the birds could have been spread and diversified so rapidly ahead of the European colonists who were said to have introduced them.

Whatever its origins, the breed was introduced to a wider world by Prof. Salvador Costello, who described the Araucana to the 1921 Poultry Congress at The Hague. His description, together with an article in *National Geographic* magazine, sparked a craze for the 'Easter Egg Chicken' which has rumbled on ever since.

One of the problems has been that the original birds from Chile came with a complex series of decorative bits-and-pieces – crests on their heads, beards under their chins and whiskery muffs on their cheeks, some with tail feathers and some 'rumpless' without tails, some with curious ear-rings of feathers springing from wart-like growths on the sides of the head and others clean-faced.

The ear-ring is a curiosity in itself. It is widely believed by devotees of the breed that it is the expression of a lethal gene, similar to the crested or dominant white genes among canaries. This dictates that when a male with ear-rings is mated to a female with ear-rings she will produce a proportion of eggs which will fail to hatch. A clean-faced bird mated to an ear-ring bird, on the other hand, will produce 50-50 of each type and no dead-in-shell chicks. A Dutch authority recently rubbished this genetic theory, so you will have to experiment for yourself. It's little wonder that the Araucana was described by Dr. William Cawley of Texas A & M University as 'poultrydom's mystery chicken'.

For half a century, enthusiasts argued about the physical appearance of the 'real' Araucana. The debate grew so heated that poultry magazines on both sides of the Atlantic banned all letters on the subject for years.

It wasn't until quite recently that breeders agreed to differ and split the Araucana into two types: the tailed bird with muff and beard, which is called the Ameraucana (often misspelt as 'Americana') in the States, and the rumpless with ear-tufts.

Some poultry folk on the Continent are still trying to keep this long-running dispute on the boil. Referring to the Ameraucana, a French website notes sternly: 'This race is not approved by the European Commission.'

Even so, the tailed and the rumpless forms are both popular in the UK. The cocks weigh about 6lb., the hens a pound less, and the breed comes in most colours. There is also a bantam, usually rumpless, with cocks weighing around 2lb. (900gm.). I recently came across a flock of wheaten-coloured rumpless Araucana hens, living at full liberty in an orchard, and they were among the most charming chickens I have ever seen.

The first standardised Araucanas in Britain were created during the 1930s in Scotland by George Malcolm, whose original stock was said to have come ashore on the Isle of Mull from a shipwrecked Chilean freighter. Malcolm's full-tailed, bearded lavenders were the basis on which most British Araucanas were founded and they certainly laid blue eggs.

While all the bickering and squabbling was going on among the hard-core Araucana enthusiasts about the appearance of the bird, more commercially-minded poultrymen realised that the Easter Egg chicken was a goldmine. It not only laid curiously coloured eggs, usually blue, sometimes greenish and khaki, but it was such an oddity that other attributes were easily grafted on, such as idea that the eggs were low in cholesterol (they are not, but the notion certainly sold eggs).

From a commercial viewpoint, the Araucana's blue-egg gene is particularly attractive because it is dominant over other colours. Mate an Araucana to a white-egg layer such as a Leghorn and you get productive hybrids which lay blue eggs. If the mating is to a brown-egg layer, you get green or khaki eggs. By careful hybridisation it has been possible to create a whole spectrum of highly-saleable colours, from turquoise green via olive to blue and even peachy-pink.

Experienced breeders tell me it is possible to tell by eye whether a bird being sold as an Araucana can actually lay blue eggs. If it does not have a pea comb – that is a tiny comb with the appearance of a raspberry – it does not have the right genetic make-up. The blue egg and the pea comb live on the same chromosome, I am told, and you cannot have one without the other.

Enterprising hybridisers also blended into their dream bird the autosexing genes which simplify the selection of dark-downed female chicks at day-old and the culling of the paler-coloured male chicks. Only future egg-layers are reared. These hybrids – part Leghorn, part Barred Plymouth Rock, part Araucana – are the Legbars which produce blue eggs for the supermarket shelves.

Although they are suddenly fashionable, the Legbars have been around for quite a long time. They trace their origins to three Araucana hens brought back from South America to Stow-on-the-Wold in 1927 by plant hunter Clarence Elliott. These were taken into a hybridisation programme at Cambridge University, where Prof. R.C. Punnett produced several autosexing breeds, including a blue-egg layer, the Cream Legbar. Whether today's Legbars are direct descendants of the Cambridge birds or a 're-mix' is not clear, but the trademarked Old Cotswold Legbar is now the best-known name in the supermarkets.

'Cotswold Legbar hens have a strong muscular frame, range further and have more stamina than most hens,' we are assured by the breeder. 'The eggs have a thicker, harder shell, a denser texture and a larger yolk than most eggs. They have a good, rich flavour so are particularly good for baking, and have become a favourite among many top chefs, including Jamie Oliver and Rick Stein.'

Pure Araucanas are widely available in the UK, in all their permutations of feathering and colour, at prices from £10 each and upwards. They are quite good layers, but if you want upwards of 200 large eggs a year, you might find it worthwhile to seek out a breeder of Cream Legbars.

Alternatively, you might buy an Araucana cockerel and some Leghorn-type pullets and breed your own Easter Egg chickens. In a few years, even if the European Commission does not approve, you could create the Olde Cornish Cream Legbar or the Original Cambridge Blueshell. The Araucana seems to inspire creativity.

Fish Pond: First the digger, then the rod

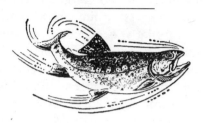

James, our neighbouring farmer, was blasting a few holes in a boggy field the other day in an attempt to encourage some wild ducks to pop in for a snack this winter. As the great clods of rushes came thumping down from the sky and the smoke drifted away, I asked our friendly local stump-blaster about fishponds. Would a stick of dynamite do the same efficient job for fish?

'Ducks are easy,' he said. 'Some explosive gives you a ten-by-ten hole, and a handful of corn does the rest. Fish are complicated.'

James had earlier been thinking about digging a trout pond, but shied off because he hates red tape and he was worred about how many office-desk wildlifers would be putting in their tuppenceworth. But his duck flighting will be a brief thrill compared with almost year-round pleasure from a pond stocked with rainbow trout, and I think I have persuaded him to face up to the paperwork.

There's no doubt that an acre of water gives better returns for the sportsman and hobby farmer than 10 times as much dry land. With a bit of thought at the planning stage – the daydreaming bit – it is even possible to have some fishing, an evening flight or two and an abundance of life, from wagtails to frogs, all from the same patch of water.

The requirements of the trout, the ducks and the 'miscellaneous' do tend to conflict with one another in a small space, so if you can make it bigger, your pond is going to be better. When we paced it out – 70 strides each way – an acre did not seem like very much, especially when you remember that trout demand fresh, cool water, clear and deep, and fishermen need room to cast, whatever direction the wind is coming from and preferably over short grass. Meanwhile, the ducks need shallow edges in which to feed, and everything else needs weeds.

If you are concentrating on the trout, and foregoing the duck shooting, you can make things easier by digging a pond with sides which slope quite steeply down to 10ft deep. Otherwise, you have to have a shallow end, a deep end and some rough edges where you can plant the weeds which technical manuals call 'emergent aquatics'.

When you have the right place for a pond, like James's marsh meadow, the project is quite straightforward: drag-line out some soil (and sell it), arrange for the overflow to drain away through a screened outlet, build a fence and pop in a few fish. The planners will be quite happy so long as you are not 'abstracting' more than 20 cubic metres a day from a stream, though they may ask you to license the pond if its demands go much above this, which is roughly equivalent to the flow of a low-pressure hosepipe.

If your pond is filling with bog-water, the bureaucrats are not going to bother you at all, but the water does need to flow or you will find that you have built a septic tank and not a fishing lake.

Dams are different, and building a big dam is high-tech stuff, starting with an impoundment licence and the services of a hydraulic engineer. There is also the problem that everything above your dam drains down into it and every-thing below is at risk if it collapses. For the 10-acre farmer, the sporting fishpond usually makes sense only if there is a low-lying patch of ground where the water table is within a couple of feet of the surface, just waiting to be scratched by dragline or JCB.

If the water is there in summer, you can assume that the pond is going to stay full all year round. If you start building a dam, you have to find out whether the soil is going to hold water – clay is perfect, chalk or sand is hopelessly porous in most cases. One of my farming cousins has a dry dam on his farm and the subject is so embarrassing that no one can mention it even now, although it is 25 years since the water ran away.

Lining a sport-fishing pond with plastic liner is a big investment and you'd be wiser to buy a rod in a local lake syndicate.

The advice we got from a fish farmer for our own syndicate pond was that if you want to catch big sporting trout you've got to buy big sporting trout – introducing 8–10 inch rainbows in spring gives foot-long fish by mid-summer and 14 inch trout by early autumn. We have tended to buy them even bigger, at one-and-a-half to two pounds for £1.50 per pound, plus a few trophy fish over 3lb. at £2.50 per pound. The fish are put in when the water is cool. We stock very lightly and throw in plenty of feed pellets.

The main losses, apart from a wipe-out when the pond overheated in a rare burst of British summer warmth, have been to predators. Trout fishermen say that no pond is safe from cormorants unless it is more than 100 miles from the sea, and a quick look at the map shows that few are so lucky. And herons are everywhere.

We invited James to come over for an evening's fishing on

our pond, suggested the pheasant-tail nymph, and a couple of clean bright fish went into the frying pan in the log cabin.

The next day his JCB was down beside the blasted duckponds, starting work on his trout lake.

Fruit Trees: Time to save the vanishing apples

You may not be able to do much to save the rhino, but any hobby farmer with an acre or two can help to save the Sack and Sugar, Onion Redstreak or Sweet Sheep's Nose – all antique apples which are vanishing from our orchards.

At one time, every farm and country house had fruit trees – apples, pears, cherries – probably with sheep and poultry on the grass under the trees. Idyllic, but now largely gone. Two-thirds of British orchards have been cut down in the past 30 years, and as they were grubbed up many of the varieties which gave us the world's widest choice of delicious fruit almost disappeared into history.

But re-creating apple orchards is not difficult, and some specialist nurseries have kept the more distinctive varieties in commerce. An apple tree is long-lived, so grafting material survives for a century or more, although many of the 6000 recorded varieties have probably gone for ever via the log-burning stove.

Several hundred of the antique apples are still around, however, and a container-grown tree, which you can pop into the ground at any season, costs £10–12, plus say £2 for carriage if you live a long way from a fruit-tree nursery. You can have it on a dwarfing rootstock for a small garden or a more vigorous one which will eventually give you 20ft trees in your orchard-meadow. Apples look right in almost any setting, especially against a wall.

Constructing a stock-resistant tree guard costs more than the young tree, but that expense is well worth it when, in a couple of seasons, you are picking the exotically-spotted red and yellow Autumn Pearmain or the wonderfully flavoured Winter Banana. Modern varieties are good too; they are frequently more disease resistant. We have been particularly delighted with Discovery – a tree presented to my wife Dorothy as a birthday present – which produces bright and tasty fruit in mid-August.

There's more to apple growing than the pleasure of eating your own apples off the tree or from cool storage, from August until April – you might also find a market for them. I particularly like an American case history, the story of John and Phyllis Kilcherman of the Leelanau Peninsula of north-western Michigan. They have very long, very cold winters in upstate Michigan, but their apple trees seem to like it.

John and Phyllis became interested in antique varieties of apple, and continued collecting until they found themselves with more than 200 different kinds of eaters and cookers in their orchard. So they printed out the descriptions and histories of the various types, boxed the fruit in sample packages, and let friends know they were for sale. The result has been sensational: they now have customers in all 50 states and even as far afield as Norway, India and Russia.

Unlike supermarket fruit – groomed to ripen at the same time, at the same weight, bruise-resistant and looking ravishing under the striplights – the old-timers are often less evenly matched but are winners for flavour and texture.

It has been said that customers eat apples with their eyes,

the reason that Red Delicious is a top seller despite a chewy and bitter skin. That variety has been described as a 'victory of style over substance', retaining its cheerful good looks long after the flavour has departed. Most supermarket apples are heading down the same eyecatching route.

You might be more tempted by Winter Banana, Strawberry Chenango, Winesap, Cornish Aromatic, d'Arcy Spice, Irish Peach, Nutmeg Pippin, Ronald's Gooseberry Pippin or Pitmaston Pine Apple. There's even a Poor Man's Profit, which seems a good starter tree for the small-scale farmer.

Your climate and soil might not suit a variety you find attractive, so it makes sense to check out what grows where. Forfar Pippin might be more suited to your short summer than the Mere de Menage, although warming temperatures are on your side and apples survive our winters very well.

Late-season apples may not mature in cool areas and very early ones are often lacking quality. You also have to keep in mind that some free-bearing apples may tend to go biennial in their fruiting pattern if you fail to prune them (or, in my own case, go biennial because the bullfinches had a bonanza spring and ate most of the flower buds).

There is a fascinating choice offered by nurseries such as

> Bernwode Plants,
> at Ludgershall,
> between Oxford and Aylesbury
> (01844.237415)
> (www.bernwodeplants.co.uk)

who list about 200 varieties, noting whether they are best for dessert, cooking or cider. Some types dry well, others are famous for their apple-sauce qualities.

What is the Next Big Thing in apples for the hobby

farmer? Well, you could keep your eye on the pink apples.
Like the pink grapefruit, they have been lurking on the
edges of the market for a long time and could suddenly
become a foodie fashion item. Look out for Surprise, a
green-skinned, pink-fleshed curiosity which was noted at
Chiswick as early as 1831, or its American successor
Hidden Rose – 'deep rose red, crisp, juicy, sugary and richly
flavoured.'

And the best apples of all time? Dr.J.M.S. Potter, one-time
director of the 3000-apple collection at Faversham, Kent,
had a list of five:

> Ashmead's Kernel,
> Ribston Pippin,
> Cox's Orange Pippin,
> St. Edmund's Pippin,
> all British in origin,
> plus one from the USA,
> American Mother.

Game Crops: Solutions for food and shelter

The first question is: why would a hobby farmer want to plant game crops? Ten acres just don't have room for any serious shooting, other than a duck flighting pond.

And the second question is bound to follow: isn't a patch of game mixture just a waste of space?

The answer in many cases is that an acre or so can often be a useful bargaining counter with a landed neighbour – and a couple of days' shooting next door is better than none at all. There is also the bonus that you can have a great show of pheasants all winter, at very little cost, and many of us really do like to see these gaudy, arrogant birds about the place, to say nothing of clouds of greenfinches and yellowhammers.

As the year moves into the 'hungry gap', the harsh weeks at the end of winter when stock and wildlife seriously feel the pinch, the shooting parties have gone home but the wild pheasants and partridges are still out there, grateful for any shelter and food you can provide. This is the right time to think about game crops. Walk the bleak fields, make the plan, sow in spring, and enjoy the results over the following years.

For someone working on a small canvas, there will probably be little choice as far as the site is concerned. There is a strip along a boundary hedge or between two paddocks which has to be the right place. If it gives your neighbour an extra drive, that's a real bonus.

The more difficult choices might lie in picking the best thing to grow. The first option to consider is the simplest, which is to plant some cover and provide the food out of a bag. A mix of hardy kale varieties such as Coleor, grown as a two-year crop, plus a few scoops of wheat thrown in daily from a nearby bin is straightforward – and it works.

If you really have only one possible site, and it looks as if your game crop might become permanent, the kale can be beefed up with some shrubs or grasses which will face the harshest weather, split the wind and stand up when the snow falls. We unrolled a few yards of mulching strip – the kind that allows rain to penetrate but prevents weed growth – and stabbed in lots of *Lonicera nitida* cuttings (each about 18 inches long) straight through the plastic and into the ground. These hardwood cuttings were spaced at 4ft intervals and took very well in March. They have since made up into 4ft to 6ft evergreen bushes, without suckering or thorns, wonderfully dry and sheltered underneath, forming the backbone of a permanent game cover strip. The plastic is still there, now covered in sand, and the pheasants scratch for corn in the dim cavern underneath the evergreen canopy.

This kind of shrub covert has the advantage that wheat thrown on top of the Lonicera falls to the ground below, where pheasants feed happily but woodpigeons are less willing to push their way in.

Another island in the kale is provided by a few clumps of perennial Reed Canary grass (*Phalaris arundinacea*), one of those plants which is wonderfully useful in the right place but a menace if you pick the wrong spot because its rhizomes spread quite fast. Its tussocks grow tall, up to 5ft in favoured damp places, and its buff-brown leafy stalks stand strongly right through until the end of the winter. It is one of the earliest grasses to break into new growth in the spring and our sheep enjoy browsing the clumps in March.

Reed Canary grass doesn't look anything special in summer; the virtues of the plant are almost entirely in its ability to form long-standing, dense cover in winter.

I found my original clumps of this grass growing on a council rubbish tip and under our tough conditions it seems to be controllable, although this 'aggressive and invasive' species is regarded as a serious pest in North America. Grow it (rather slowly) from seed, or split up some clumps from a waste-land site. Pheasants find Reed Canary grass very attractive, it probably reminds them of their Asian marsh homelands. In fact, one of its disadvantages may be that they jug among the canary grass tussocks at night rather than flying up to a more fox-proof roost. (Seed from British Seed Houses, 0117 982 3691.

I have been thinking of adding some clumps of Pampas grass (*Cortadera selloana*) to our game strips because it has been so successful in the pheasant pens. This also stands up well in winter and there are few cosier places in a blizzard than the wind-proof lair at the base of a big pampas clump. A farm college near here had an experimental plot of this wonderful grass for several years. It stood in the middle of stone-dyked fields, all grazed bare, and it was a magnet for pheasants. As fodder, the razor-edged grass seems to have failed the test, unfortunately, and has now been ploughed in.

However, it's worth experimenting with pampas grass as cover. Look out for the lusher, greener clones. The local garden version, here in the North, is notably meaner, slower and duller, with a narrow leaf. It may be hardier, but I doubt it.

Despite the disadvantage of their cutting-edge leaves, these giant grasses make superb year-round cover and if they get too rampant, you can burn off the old growth. Just remember to wear thick leggings and gloves when you wade into your pampas patch, and a mask too if you cut paths through it with a ride-on mower.

Garden fashionistas have been very unkind about the suburban Pampas grass, standing proudly in the centre of so many front lawns. Forget their sneers and take a fresh look, it really is a fine species. Buy young plants at £8 for 10 from Parker's Wholesale, 452 Chester Road, Manchester M16 9HL (0161 872 3517). The little plants I put in my pheasant aviaries last spring are now 3ft high, quite bulky at the base and making useful shelter. At the end of the coming summer they will be green giants, so give them plenty of space.

And while we are on the subject of plants in fashion, the species to sow if you want to be seen to be keeping up with the trends (and none of these even rated a mention when I was putting together a booklet about Game and Shooting Crops for the Eley Game Advisory Station in the 1960s) are Quinoa, Triticale and Phacelia.

The first two are seed providers, retaining some feed value into January and even beyond, and blue-flowered Phacelia is a very fast growing annual which will give the pigeons something to do until your other crops get going – the birds can be very hard on small patches of kale.

The anti-vermin recommendation for small patches of kale is horticultural fleece. It keeps the soil warm and moist while protecting the emerging plants from woodpigeons. You pick it up when the plants are about 6in. high.

Although this costs £150 to £200 for an acre, the fleece can be re-used for several years.

Quinoa – which is pronounced 'keen-wah' if you want to be strictly correct, although everyone seems to call it 'kwin-oh-ah' in this country – is a tall (3–6ft) Andean relative of our familiar weed, Fat Hen. The hype surrounding it is considerable and it comes to us as 'the sacred mother grain of the Incas'.

The seed is high in protein, with the crop maturing at 100 days and capable of producing a ton to the acre under British conditions. It is a slow starter and its major problem seems to be that it is easily overwhelmed by weeds. However, once you get it going, a Quinoa and kale mix is currently agreed to be the best game crop virtually everywhere in Britain, although some seedsmen would add a dash of Triticale to the recipe.

Triticale (*X Tritico secale*) is a wheat-rye hybrid which first became available as a commercial crop about 1970. Since that time, the world acreage has boomed to more than 500 million acres because Triticale really does produce the goods on poor soils and under tough climatic conditions – just what you need for a game crop. Again it mixes well with kale grown as cover, although it does need to be sown quite early in the season, by April, to be productive. It has the useful ability to regrow strongly after rabbit damage.

So the advice to the hobby farmer is, go for it. Whether you opt for kale and a bucket, or a complex commercial game mix at £30 or £40 an acre, you can do nothing but good by giving food and shelter to the birds.

Geese: Turning grass into free-range meat

Geese are one of the most trouble-free and traditional ways of turning grass into meat. So long as the hobby farmer's holding is not too small, this currently rather unfashionable domestic bird can produce a tasty 'green goose' straight off the meadow by Michaelmas or early in October.

Free range has always been the essence of goose-keeping. The birds are quite noisy, so they are not ideal for a pen under your bedroom window, and they produce large quantities of muck, as everyone who has walked around London's park lakes knows very well, so they need lots of space if they are not to foul their grazing. But if you have an acre or two of green meadow grass, a small pond and an electric fence to scare off the foxes, keeping a flock of geese is a worthwhile project.

One of the birds' greatest virtues is that they are truly hardy. They are happy with anything that the British climate can throw at them. Rain, frost, ice, gales, fog, sleet – they are all lovely weather for geese. And they don't seem to mind the occasional heatwave so long as you can give them some shade.

The advantages of goose-keeping should please any small-scale stockman. They are easy to breed, rear and feed, unfailingly healthy under free-range conditions. They need only a simple night shelter, they are very long-lived, and producers of excellent meat.

The drawbacks are their need for space, vulnerability to foxes and, with some breeds, a tendency to be a bit too aggressive in defending their mates, although this can be put under their virtues if you call it 'good watchdog skills'.

The custom in the past was always to breed and feed geese for maximum fat on the dressed bird as presented on the poulterer's counter, and this led to their decline, which can be dated back to the beginning of the 20th century. Their replacement as the traditional Christmas roast was the much less fatty turkey, which also responded readily to mass-production methods.

Young geese off the meadow do not have nearly as much fat as the supermarket version, so do not be tempted to test your enthusiasm for roast goose until you can buy the right bird. A 'green', grass-fed goose is much more like a wild pinkfoot than a greasy duckling.

There has never been much effort by large commercial breeders to produce a more market-friendly version of the goose because it is not a very commercial form of poultry under UK conditions and once you get beyond the 10–20 bird scale, it does not make much sense because of low egg production (perhaps 30 eggs a year from the medium-sized breeds), seasonal breeding, and low fertility under intensive conditions.

In countries where there is still room for extensive grazing on marshes and alongside waterways, with cheap labour to herd the birds, the goose maintains its popularity. And on the British smallholding with appropriate grazing, middle-sized geese are a great addition to the stock. Very few people keep them in large numbers in this country except those who have found a niche market for high-quality breeding stock.

A neighbour of mine, who tries to stay ahead of fashion with his poultry by breeding the types which are booming in popularity and quickly sidelining the rest, tells me that the big white geese have been replaced by 'ordinary' breeds, like those believed to be nearest to the geese of the old farmyard and meadow, notably such varieties as the American Pilgrim, Brecon Buff, and West of England.

The West of England is a standardisation of the most basic kind of English goose, the one always pictured standing beside the pond on the village green, with mainly white ganders and grey pied female geese. It took centuries for the penny to drop with farmers that their geese had a natural tendency to produce sex-linked stock in which the males were white and the females had a mix of grey and white in the plumage.

In fact, an American waterfowl enthusiast, Oscar Grow of Iowa, has a good claim to being the originator of a truly auto-sexing goose breed, which he established before 1935 and named 'Pilgrim'. The origins of the name are obscure, with most poultry historians assuming it came from a link with the Pilgrim Fathers and early settlers, although Oscar Grow himself said it referred to his own family's pilgrimage across the Mid-West in the late 1920s during the Depression years.

Whatever its origins, the name for this improved English goose was being used by the mid-1930s and the breed was standardised in 1939. The auto-sexing really does work, with day-old male goslings having pink beaks and the females dark beaks. As adults, all Pilgrim ganders have blue eyes and the geese are dark-eyed, in addition to the plumage differences.

The Pilgrim is officially a light breed, alongside the upstanding, noisy Chinese (whose white form is the most elegant of park geese) and the neat white Roman, although a 15–16lb gander would not seem lightweight to most people. Average males weigh about 14lb and females go a pound or two less. They grow fast, with goslings putting on nearly a pound a week on grass with only a handful of poultry rearing pellets each day. When you watch these waterfowl growing like weeds in high summer, you can see why they were so popular with cottagers and small-scale farmers in the past.

Like sheep, geese have complex emotions even if they do not have much analytical intelligence, and they respond to good husbandry. Evil tempered ganders give some breeds a bad name, but the lightweight and medium types are not difficult to live with when they are handled gently, and it's worth getting it right because a gander may be around for a dozen years or more, so you don't want to create a monster who forces visitors to pick up a broom every time they come through the yard.

Even respected poultry writers like Lewis Wright, who produced the classic books on fowls and waterfowl, believed that a gander could break your arm if he really lost his temper, so in these litigious times it is not wise to have such a bird loose on your hobby farm.

You can see why one American breeder emphasises that his Pilgrims are 'exceptionally calm, sweet natured (even personable) and self-sufficient. They are quiet and docile. Although they are much less aggressive than other breeds of geese, they can become protective of their newly-hatched goslings.' And he adds: 'Temperamentally, only calm and sweet natured Pilgrims should be bred, since good temperament is a part of Pilgrim breed character.' Like many other small-scale breeders he keeps his birds in pairs or trios, although one gander can be mated to as many as three or even five geese. They sit on their own eggs (28–30 days) and rear their own goslings.

The cottagers' habit in England was to give 13 eggs to the goose to hatch herself, with any extras being put under broody hens, but this is not very efficient because even a big fowl has difficulty with more than three goose eggs and she needs help to turn those, so you might find it easier to eat the surplus in the form of tasty omelettes.

Famous research by Dr. Konrad Lorenz in Austria in the 1930s showed how geese become imprinted on foster parents within a few hours of hatching and this is one of the problems if you use incubators and artificial rearing systems. At least parent-hatched goslings do know they are geese. The system might seem rather inefficient, but it avoids the rearer finishing up with the feathered equivalent of a bottle-reared ram lamb.

The Milking Goat: Fascinating, but you need lots of time

The milking goat was the perfect animal for the old-fashioned, stay-at-home smallholder. Today's hobby farmer, whose life is probably more complex, may find that keeping goats is fascinating but needs quite a lot of planning.

Goat husbandry is basically very easy. Everywhere from China to Chile there are simple folk keeping goats successfully, millions of them, often in the harshest of landscapes. Long after the cattle and most of the sheep have died of hardship, the herds of goats are still there, and they produce the world's most popular milk.

The reason that peasants' goats do so well is that they have time to devote to them. The animals need someone's time, even if it's only the attentions of a ten-year-old goatherd with a big stick.

The heyday of the milking goat in this country was in the years just after World War II, when cow's milk was rationed, labour was still quite cheap and the nation's goats were showing the benefits of considerable upgrading in the 1930s. Improved alpine breeds had been imported and milk yields had increased dramatically when compared with the amounts previously produced by our native breeds. Everyone with a few rough acres seemed to have a couple of milking Saanens.

Gradually goat-keepers polarised into two groups: the serious ones, who keep large numbers semi-intensively in yards and sell hygienically-approved milk and interesting cheese, and the pet-keepers, who keep three or four on marginal land and use the milk themselves.

On a small scale, goats are great fun. They are hugely rewarding, even if they never make money. They are active, interested and endlessly fascinating. However, they are not for anyone who wants to shut up the shop and go away for several weeks to Davos or Barbados. 'Feed them and check that they've got water' may be fine for sheep and poultry, but the casual stand-in will not be able to cope with milking goats.

Fences and boundaries are a challenge to the goat. She will be out, over or under any fence that has not been designed with her in mind, and she might be quite badly damaged in the process if the barrier is a standard 3ft 6in wire fence with a top strand of barbed wire. A 4ft fence of Rylock sheep mesh is a much better bet, but even that might need a strand of electric wire as well.

She also needs shelter because goats are thin-skinned and chill readily. The modern milking goat's protection against the cold does not come from a heavy insulating coat on the outside but from the massive central-heating system of her fermenting gut. A 140lb. goat has about 50lb. of vegetation cooking in her belly. This source of heat works most efficiently if she has access to rough grazing or has plenty of hay, and to keep the system working at full blast a milking goat eats two or three times as much roughage as a sheep of the same bodyweight.

Nettles and thistles, briars and brambles all provide heat and energy for the browsing goat. The ideal 'pasture' is not a grassy paddock but a cut-over patch of woodland with a couple of years of re-growth, which will support one animal per acre or even more if there is some hay available in winter.

To achieve nearly a gallon of milk every day from your milking nanny over a long lactation of two years, you have to give her a richer diet, with added concentrates as well as hay. As a result, the central-heating loses some of its efficiency, which is based on the fermentation of the fibre, and the high-yielder feels the cold. A shed is essential, preferably one with some insulation, a warm floor and plenty of bedding.

The serious goat-keeper has milking machines in a milking parlour. The hobbyist has a bucket and a quiet corner in the goat house. Singing or talking quietly to reassure the nanny that all is well with the world, about four pints can be drawn by hand in five minutes. Goats are milked twice a day – the main reason why the goat-keeper has to be on site all the time. A stand-in who can milk goats is worth his or her weight in gold.

The female goat is mated in the autumn and produces her kids, usually twins, after five months. Unlike a cow, she does not automatically dry off after ten months or so of production and she can be milked through for about two years before carrying another kid. This is where she scores as a peasant milk provider because her smaller quantities are produced over a long period on a simple, rough diet.

The milk differs considerably from cow's milk, with the fat being distributed throughout in very small globules. Before the days of homogenised cow's milk, which is now the supermarket standard, the goat's milk had the great advantage of being suitable for freezing. And it is more readily digested by many young animals, from piglets to puppies.

Goats should be kept in company, preferably with other goats, or they can become very noisy and demand endless attention. They will thrive on pasture land, so long as they can find shelter, and they also do well in concrete-floored yards with plenty of bedding. Wherever they are kept they need salt-mineral licks and an ample supply of clean water.

Tethering on open pasture might seem like a simple way of keeping goats under control but they don't like being tied up and they soon foul the land. They are also quite tender in wet or cold weather so they suffer (and their milk yield falls) if they are staked out in the fields.

The goat's lifespan is 8–12 years and growing-on young goatlings is one of the pleasures of keeping them – they are among the most delightful of young animals, endlessly playful.

Female goats do not smell. However, the buck or billy definitely does, particularly during the autumn breeding season. The odour has a throat-catching quality that is not easily shaken off. It is recommended that you should work with a male goat only in waterproofs which can be discarded before returning to civilised company! The smell contaminates any milk within a considerable distance, so billies (which are also quite dangerous) tend to be kept by a few specialists and female goats visit them for service. Artificial insemination is also available.

The most popular milking breeds in Britain are based on alpine importations made in the middle of the last century – British Alpine (black and white), British Saanen (white) and British Toggenburg (brown and white). The lop-eared, roman-nosed Anglo-Nubian comes in various brindled and pied colours. This breed has Eastern blood, and produces milk with even higher butterfat than the alpines, but at slightly lower yields.

Male kids of dairy breeds are almost invariably destroyed at birth or castrated. The neutered buck is quite an amiable creature, without the problems of stink and aggression, and is very easy to keep, needing only some grazing or browsing and an ample quantity of hay all year round. In exchange, a hand-reared wether goat can become a surprisingly useful pack animal.

On a couple of memorable expeditions, I have travelled in the Rocky Mountains exploring trails with the support of pack goats carrying the personal gear and lunch packs. The big wethers were happy with up to 40lb on their pack saddles, finding their own food (plus titbits) and not even needing a leading rein. They seemed to enjoy it as much as we did and led the way up the mountain paths.

Pack goats could be an interesting if slightly eccentric enterprise for the hobby farmer in many of Britain's hill-walking areas. A few hand-reared wethers would be amusing lunch-carriers on trips for outdoorsmen whose knees are beginning to feel a little creaky or for families with children. Import a few pack saddles ($165 each) or make your own, and you're in business. There is little competition so far.

Pack goats have been a familiar sight in the Himalayas and other mountainous regions for thousands of years and they are much cheaper to run than ponies – besides being better company. They will easily cover 15 miles a day, limited mainly by the physical condition of their human companions, and they can be taken to the starting point of the day's trek in a small trailer or covered pickup.

# Grapes: A little glass helps the vine

The Black Hamburgh vine in our Victorian greenhouse is heavy each autumn with more than 140 bunches of grapes, proving that the most dramatic influence on climate change is a simple sheet of glass.

Somewhere north of the Severn-Wash line, outdoor grapes become a chancey enterprise. With the help of a south-facing wall, that range can be extended up the map by about 200 miles.

However, if you put some glass in front of the wall, it is possible to grow good grapes anywhere in the British Isles. At one time the famous Kippen Vine, near Stirling, was the biggest vine under glass anywhere in the world, producing 2000 bunches of Gros Colmar grapes every year.

You don't need heat, just glass – or in the case of our Black Hamburgh, some sheets of clear corrugated plastic.

The story of this vine makes the point. It has been in the walled garden since the 1860s, planted in a lean-to vine house which was heated with an elaborate system of hot water pipes fed from a coal-fired boiler. As the gardens collapsed into decay, the glazed roof fell in.

The vine, whose main trunk is as thick as a man's thigh, did not die. But it did not produce any grapes either.

Nobody had seen a grape on it for years. In fact, no one was sure whether it was supposed to produce black grapes or white. It just survived as a tangle of intertwined shoots among the heaps of shattered glass and smashed wooden framing.

When I cleared out the debris and re-covered the roof with corrugated plastic sheets, the vine grew riotously. Within a few months the vinery was filled with an almost sub-tropical abundance of leaves and shoots on the wires under the roof. But there were no flowers and no grapes.

And then, after a year under cover and some hard pruning, it bloomed. And it set fruit – dozens and dozens of bunches of beautiful table grapes, ripening in October.

The lesson is that vines are tough plants in winter but need some help in summer. And they are wonderfully productive when you concentrate their minds with the secateurs. They do need two hot summers in succession, the first to ripen the wood and set buds, the second to grow the fruit – and under glass every summer is hot.

Even outdoors, the summers have mostly been kind lately and English vine growers have been on a high since the 2003 grape harvest. It was generally agreed to be the best ever. There are nearly 400 British commercial vineyards, producing about 2.5 million bottles of (mainly) white wine, and their owners really enjoyed the fruits of that long hot summer.

If you would like to have some home-grown grapes and maybe your own wine, the place to go in September for inspiration is the national collection of hardy outdoor vines, maintained by Brian Edwards at Sunnybank Vine Nurseries in Herefordshire. There are free open days each autumn, with about 300 grape varieties on show, both in the open and under glass.

The Sunnybank website gives you the flavour of what's growing in the national collection and what's on sale. For instance, among the black grapes, there is **Gagarin Blue:** 'Dessert/Wine grape. An outdoor dessert grape of fine quality. Early mid-season, big crops of big bunches of big loose berries. A Russian hybrid. Recommended for amateur use. Decent wine. It is a purple/black, well-flavoured dessert grape from Russia, named after the astronaut and reputedly smuggled here during the Cold War via an embassy in exchange for a box of Biros.'

Or perhaps you might be intrigued by **Seibel 13053:** 'Wine grape. Mid-season. Ultra reliable disease resistant hybrid giving big crops of strongly flavoured small grapes. It makes a reasonable red, quite good rose or excellent port style wine. Strongly recommended for the amateur. Small commercial plantings for rose would make economic sense. Very attractive vine, red shoots and bronze young foliage' (Some books list this under the name Cascade).

And among the black grapes there is also **Triomphe d'Alsace:** 'Wine grape. Mid-early, very vigorous, disease free hybrid. Good crops when the set is good, can make excellent fruity wine. THE prize winning variety in the West, increasingly popular. High sugars, good colour, needs no faking to make good wines. It can be grown organically, and for an amateur, makes a good cover vine. Highly recommended.'

There are scores of whites too, including **Seyval Blanc:** 'Wine or Dessert grape. Formerly Seyve Villard 5/276. Mid-late, a disease resistant hybrid.Very heavy crops (known as 'save-all' among West Country vignerons). Fine crisp dry wine, a winner of top awards for English wine. Also good in blends. Good and easy dessert on warm walls or under glass. Can be grown organically. A number of strains exist.

High acids, low sugars, needs careful winemaking. An easy vine and thoroughly recommended.'

American sources say that when Seyval's grapes are harvested at optimal maturity, wines have attractive aromas of grass, hay, and melon. Seyval is also praised as 'a heavy cropper and easy to grow' in the best small handbook for British smallholders, Grapes: Indoors and Out, by Baker and Waite (see appendix).

If you are looking for a reliable and low-cost way to grow a single vine under unheated glass, you might like to construct 'the curate's vinery', as described in the Baker and Waite book. This is a simple and cunning device, which gives the advantages of a wall with nothing more elaborate than a few paving stones laid on the ground.

In essence this 'vinery' is a large cloche – about 6–8ft long – built on top of paving slabs with ventilation at the sides and top. The vine is planted outside the cloche and grown inside along a wire about 6in above ground level. The grapes develop lying on the hot stone surface.

The curate's vinery is a good home for your Black Hamburgh, which is still one of the most popular varieties under glass, and with average luck you should ripen at least half a dozen bunches in the autumn. Make sure the mice and sparrows can't get in.

The humble curate may have been thinking about economy, but Victorian gardeners were given to flamboyant and expensive gestures, encouraged by their employers, and they perfected several techniques for producing dramatic 'table vines'. A 3ft-high, pot-grown vine brought to the dinner table, laden with three or four bunches of fine black grapes, scores a lot of points as a finale for a supper party.

To grow a table vine, a long whippy rod of a mature glasshouse vine is led up through the drainage hole in a 9in.

flower pot, and then staked and pruned so that there are three or four mature buds above the rim of the pot. During the summer, the shoot remains attached to the parent vine but also grows its own roots in the pot. In early autumn, when the bunches are ripe but before the leaves start to yellow and fall, the rod is cut at the base of the pot and taken to the table as a self-supporting vine.

Warmth and sunshine produce sweet grapes and in the South of England the climate has recently been kind enough to give a harvest in most seasons even where the vines are grown in the open on post-and-wire supports without the benefit of glass or wall. For the southern smallholder with a south-sloping field, sheltered from the wind, there is always the temptation to plant a vineyard.

Even if you already own the land, it has been estimated that the cost of establishing a vineyard is about £5000 an acre, and the profit on a £5 bottle of wine was reckoned in 1999 to be about 27p, so the margins are slender. But the good news is that English wine is no longer a joke: sunny summers and the introduction of hybrid vines have combined to improve the product enormously.

The hybrids have a parentage which has blended the virtues of the familiar Vitis vinifera, the progenitor of European commercial grapes, with North American species, to achieve disease resistance and earlier maturity.

If you want to start an argument among English vine growers and makers of wine, just raise the subject of hybrids. These are agreed to be vigorous, prolific and disease resistant. Their critics, however, say that some of them introduce 'exotic overtones' into wine. The description ranges from 'strawberry flavoured' to 'foxy' or Seyval's 'grass, hay and melon' – but you might like it, especially if it's your own vintage.

The grape vine is a wonderfully adaptable plant and strikes readily from cuttings. Grow it indoors or out, up a wall or confined to the curate's cloche, show it off as a standard in the middle of a lawn or use a hardy and rampant hybrid such as Brant to smother an ugly shed. Young pot-grown vines are very cheap – as little as £4 or £5 will buy you a good one – so have fun. The end product, the home-grown grape, is the finest of fruit.

Grass Carp: The stuff of rural legend

Nearly 40 years ago, the Ministry of Agriculture imported 2000 grass carp from a state fish farm in Hungary and released them into the waters of the Fens. The plan was to study the effectiveness of this Chinese fish in controlling unwanted water-weed.

Liverpool University also climbed on to the biological-control bandwagon and put experimental carp into netted-off sections of the Lancaster canal. Within a few years the grass carp was everywhere, from Buchan to Bodmin, introduced by lake-owners who wanted it to hoover up their aquatic weeds.

Since then, the grass carp has achieved the status of an almost mythological monster fish – one which is said to grow to vast size, eating everything in the water and sometimes beyond, leaving devastation in its wake. Stories abound of big grass carp going berserk and even jumping into boats.

Meanwhile, over the years, as these Oriental carp became the stuff of rural legend, the Ministry of Agriculture not only changed its name but also changed its policies. The

Good Idea of controlling waterweed with browsing fish came to be seen as a *Very Bad Idea* and the grass carp was put on the list of exotics which must not be bought, sold or released into the wild without a licence.

Unfortunately, so many had been introduced into British lakes by then that an open licence had to be issued so that all existing carp stocks could be legalised. If you have grass carp in a your smallholding pond – a private lake of less than an acre,with no outlets to the wild – you are probably covered by this general licence.

Today the grass carp continues to be treated by official-dom as an unwanted alien. Among anglers, it is seen as a coarse-fishing curiosity, elusive and shy but putting up a good fight when hooked. And despite decades of research, proprietors of weed-infested lakes on the hobby farm are still trying to make up their minds about its value as a devourer of trash vegetation.

All the grass carp in our waters have been released by anglers or water managers. The species does not seem to breed in any British lake and, like the rainbow trout, is dependent on captive-bred stock, which makes its unpopularity with the authorities rather surprising – without fish-farm re-stocking, it would gradually fade away.

The grass carp (*Ctenopharyngodon idella*) has the appearance of an enlarged version of our native chub, and its original home is in the slow-moving rivers of lowland China, where it has been an important food fish for thousands of years. Throughout South-east Asia it is farmed intensively, the fish being penned in bamboo enclosures – often under the floors of stilt-supported houses over the river – and fed on grass clippings. In sub-tropical conditions, with average summer temperatures in the water reaching 68-70°F (20°C) or above, the grass carp grows fast and breeds prodigiously.

Even in central European fisheries it has been recorded as achieving 50 pounds in weight and more than a yard in length within 10 years, although in British lakes it is more likely to grow at the more modest and unprofitable rate of about a pound a year and it can never work up any enthusiasm for breeding in cool, still waters. In the wild, it needs a flow rate of 2–5 ft. per second in a warm river before it is prompted to start egg-laying.

Very slowly, our British grass carp do reach considerable size. The record is a fish taken from RMC Angling's Horton Church Lake, Berkshire, by Kevin Ballard, of East Grinstead, which weighed 39lb.2oz. For the cognoscenti, it is noted that 'Kevin presented a Nutrabaits cranberry Big Fish Mix boilie on a stiff hinged pop-up rig over a bed of Dynamite Baits Frenzied Hemp and crushed tigers.'

Well, it must have made a change from duckweed.

American anglers recommend floating bagels and ground bait mixed with black molasses. Some very big English fish have been tempted by dog biscuits. The grass carp may be shy, but it obviously has a pioneering taste in exotic foods.

The grass carp in the hand is very different from the bag-of-sludge outline of the common carp, as so often photo-graphed in the arms of a smiling coarse-fishing enthusiast. It is long, streamlined, with a small head and low-set eyes almost in line with the mouth, and the colour is a gold or silvery-sheened bronze over an attractive reticulated pattern of scales.

There is also an albino form, popular as a pond fish and probably for sale in the aquatic section of your local garden centre, where you can expect to pay about £2.50 for a two-inch fish.

The drawback of the grass carp in a small garden pond is that it does not know when to stop growing. Many species of fish become stunted if their surroundings do not allow them to develop fully, but *C.idella* is adapted to big rivers and goes on getting larger and larger until it runs out of food or oxygen. Along the way, it can leave other species without food and shelter, although it is not directly aggressive to other fish.

In hot weather, grass carp do eat a lot of greenstuff. When water temperatures reach nearly 80°F (say 25°C) they have been recorded as devouring their own bodyweight every day in duckweed, blanket weed, Canadian pondweed, pennywort and other pest plants, which are all grist to the carp's mill.

In some warm countries, big fish have been known to consume nearly a hundredweight of aquatic plants daily, and sometimes they lurch out of the water to grab overhanging greenery, which is dramatic. They get quite frisky, and there are authentic records from America of grass carp jumping into small boats. But if the temperature drops, their appetite and activity rapidly fade.

The stocking rate if you are tempted to use grass carp as a biological form of lake weed control on the smallholding is usually set at about 5–15 fish per acre, depending on the amount of weed infestation, water depth and temperature. Small fish do not have much effect, so the introduction of quite large carp is recommended. Expect to pay about £25 each for foot-long fish.

What happens if they eat all the plants and you want to get rid of the carp?

I put this question recently to a lake owner in the American state of Georgia as we looked at several huge grass carp basking in the shallows under the loblolly pines with their hanging trails of Spanish moss. It would take a lot of time and skill to catch them with rod and line, I suggested.

'Have you ever heard the expression: shooting fish in a barrel?' he asked. 'That's the answer. They're hard to catch, but not difficult to get rid of.'

Although the various carp, including grass carp, are the world's most popular food fish, they are held in low esteem in America, where they are called trash fish and accused of rooting out all the waterside plants which provide waterfowl feed. Duck hunting is much bigger business than coarse fishing. As a result, carp are more often used for bowhunting target practice than sought out by specialist anglers.

There are strict laws in most states to prevent fertile grass carp from being released and dealers can supply only sterile triploid fish. The same rules apply in parts of mainland Europe.

In Britain, the grass carp is just one of the dozen or so non-native fish which are now found widely in our rivers and ponds. Some, such as the common carp, which was imported to monastery stewponds in medieval times, are today accepted as virtual natives. Others, such as the predatory zander, are seen as problem species.

In the middle, the grass carp remains a puzzle and no one is sure where its future lies. A few anglers call it 'this fantastic fish,' delight in its challenging shyness and sensitive mouth, and admire the fact that it 'goes mental' when hooked. They urge the introduction of this 'elusive and hard-fighting' trophy to more lakes and reservoirs.

However, environmental pessimists believe it could be a source of problems if the summer climate gets hotter and it becomes a breeding member of our river fauna, eating the natives out of house and home.

The Environmental Agency said recently that angling, which is Britain's favourite pastime, is now threatened by global warming. Wetter winters and warmer summers, say the doomsters, could so upset the life-cycle of our freshwater fish that they are likely to be threatened with extinction. In which case the mysterious, warmth-loving grass carp could have a brilliant future.

Guinea Fowl: Hardy, noisy and great on the table

Guinea fowl are currently enjoying a burst of popularity on both sides of the Atlantic – in Europe because they are good to eat, in America because they eat garden pests.

In crowded Britain, this fascinating gamebird can claim fewer enthusiastic supporters because, despite all its virtues, it has one big drawback. Like the peacock, it is noisy.

Free-ranging flocks of guineas like to talk to one another all the time. The hens have a far-carrying 'go back, go back' call, the cock birds have a one-syllable screech, and when alarmed the whole flock will explode into a volley of sound rather like machine-gun fire as they fly up into the trees. In densely populated England, with nearly 1000 people to the square mile, there is usually someone within earshot who complains.

For the hobby farmer whose acres are not bordered by too many sensitive neighbours, however, guineas are great birds to have around. And their calls may even bring back memories of African safaris and mornings in the bush. They are brilliant guard dogs, with the added plus of being low-maintenance pest controllers.

For those who might believe this is not enough, there is also the thought of Delia Smith's 'Guinea fowl baked with

30 cloves of garlic', which she describes as 'utterly sublime', and the knowledge that Continental gourmets are so enamoured of the pintade that 54 million are bred for the table every year in France. Production of guinea fowl in Europe has been increasing steadily – it was up 5 per cent last year – and nearly half of the table birds sold in the UK are imported from the Continent.

Although it might seem to be a reluctant member of our domestic poultry yard, preferring semi-liberty to the safety of the wire pen and tending to form pairs rather than more economical harems of one cock and several hens, the guinea fowl has actually been domesticated twice, first by the ancient Egyptians and later in West Africa.

Egyptian tomb paintings show that their domestic guineas were based on an East African race of the wild helmeted guinea fowl rather than the West African type of our familiar bird, which was apparently brought to Europe by the Greeks and Romans and later by Portuguese explorers.

Despite those equatorial origins, the guinea fowl is remarkably hardy. In fact, several authorities on poultry consider it the hardiest of all, apart from waterfowl, and certainly I have had a flock which wintered out of doors in the foothills of the Cairngorms, roosting on the top branches of old apple trees. At night they often had snow on their backs, but they twittered away quite happily in the moonlight and flew down every morning as if it was dawn on the slopes of Kilimanjaro.

With early training, it is possible to persuade the birds to roost in the rafters of a large barn. However, their preference is for a tree roost and once up there in the branches they are safe from predators – too big for cats or owls to attack and far above the reach of foxes and dogs.

Guinea eggs hatch after an incubation period of 26–28

days and the gratifyingly vigorous chicks or 'keets' need to be kept out of the rain and wet grass for the first 2–3 weeks, but after that age they are tough. Turkey or pheasant starter and grower rations are fine.

The keets go up to roost enthusiastically at about 6–7 weeks; they need very little encouragement. Until that point they are vulnerable to foxes – which is the reason England is not overrrun with guineas – so for the small-scale breeder the young birds are best reared under broody hens in movable pens on grass, then released to thrive at semi-liberty. A broody turkey hen is even better, if you have one, because she will keep them together until the following spring, whereas a bantam will ignore them after 50 days. The turkey may also persuade them to roost inside.

Artificial rearing is straightforward too, with a starting temperature under the lamp of about 37°C (say 96–97°F) reducing at 5°F each week. Anyone with experience of turkeys or pheasants will find guineas very easy.

Rearing from day-old is the only sure way to start a flock of guinea fowl because birds bought as adults are very difficult to settle. If they are released, they disappear. If they are kept penned up, they may be reluctant to breed.

On free range, guinea hens make their own nests in the long grass. However, they are nervous sitters even though guarded by their mate, and will abandon most early nests, leaving their owner to hunt for the eggs. On average, guineas will lay about 30 eggs before sitting.

The wet grass and the foxes take their toll of keets hatched in the fields and woods, so finding the eggs is essential. For this reason, rearing guineas is one of the most haphazard of livestock enterprises on the hobby farm. It is possible to work with them under more confined conditions – they couldn't produce those 54 million birds in France if that wasn't the case – but their other virtues are lost if they are treated as penned pheasants, mated six hens to the cock and fed on turkey rations.

The eggs are very thick-shelled, which makes them difficult to candle when checking fertility, and they offer the unusual bonus of being sure-fire winners of any Easter egg-rolling contest. They never break. The flavour is rich and they can replace hens' eggs in many recipes.

Around the smallholding, guinea fowl are decorative as well as useful, and score highly on personality, even though their naked heads may not appeal to everyone when they are viewed in close-up. As one farmer puts it: 'Both their looks and their voices are like the kilted bagpiper, wonderful at a distance!'

Apart from the wild-type plumage – lavender grey on the neck, with white-spotted black feathers on the curiously turtle-backed body – guinea fowl come in about two dozen different colours across a spectrum of buffy browns and greyish blues. These range from chocolate and pewter to purple and lavender, and there are also pure white and pied variations. Legislation governing the shipping of poultry, which has almost wiped out the conservation of rare pheasants in captivity in this country, also has its effect on guinea fowl, and unfortunately it is impossible to recommend a breeder of these colours who can send stock around the country. Local sources of supply have to be found in markets or the small ads.

Distinguishing the sexes of guineas is difficult when they are young. As they mature, the males develop a larger bony casque on the crown of the head and they also become leggier and slimmer than their mates, but the earliest certain difference can be seen at 12–15 weeks by examining the wattles on the side of the face, noticeably thicker at the

front edge in males than in females. Adults are most readily
sexed by their voices – only the hens have the 'go back, go
back' call.

The current boom in popularity of guinea fowl in
America is based not on their looks, or even on their
gourmet quality, but on the birds' usefulness in controlling
ticks. Much of rural North America is almost paranoid
about tick-borne ailments, especially Lyme disease, and the
guineas are known to be very efficient in keeping tick
numbers down. The birds also dispose of a wide range of
other pests, from caterpillars (including the ones which
most birds refuse to eat) to field voles.

Unlike hens, they prefer to pick at eye level rather than
scratch in the soil, which means they create much less
damage in the garden, although they do like to dig a dust
bath somewhere sunny, secluded and dry. In summer they
may get as much as 90 per cent of their food in the fields,
but they always come home for a top-up feed of wheat in
the late afternoon.

As a sporting bird, guineas have been a disappointment
outside their homelands of central and southern Africa.
Rather than fly long distances, they will usually take to the
treetops when flushed by a dog, and sporting writers of the
Victorian era noted rather disparagingly that 'native guns'
take advantage of this habit to bag a few for the pot.

In addition to being an unwilling high flier, the guinea
cock is notoriously aggressive towards pheasants, which are
completely demoralised by his thuggish style of street
fighting, and keepers who used to put a few young guineas
to covert with their poults to encourage the pheasants to
roost, always had to catch them up soon afterwards to
prevent the Africans from taking over the feed rides.

Around the farmyard, guinea males rush at other birds,

their back feathers raised, and their hit-and-run tactics ensure that they are always cock-of-the-walk. However, they seldom become aggressive towards people.

Hedges: Defining the feel of the lifestyler's farm

Decisions about creating hedges for the smallholding or garden are often made during the winter, at a time of year when evergreens appear to be the obvious choice. The charm of greenery and the value of shelter outweigh most other considerations when the days are dull and an east wind is cutting across the paddock.

Even during the summer, the evergreen Leyland cypress might be your choice for no better reason than because it is available all year round from nurseries as a potted plant, grown from cuttings, ready to pop into the ground with minimum fuss.

But there's a lot to be said for having a look at your local hedges during the spring and summer, just to get a few ideas about the possibilities of field boundaries which have a more varied appeal than the evergreens, with flowers and berries in season.

Fashion in farm hedges has not changed much over the past 300 years. Hawthorn – often called quickthorn in this context – is still the right choice for most soils and most climates, especially if the fields will be holding livestock. The whips are easy to plant and they do survive.

However, hawthorn is increasingly seen simply as the

main ingredient in a hedge-planting mix of species rather than as a stand-alone boundary plant. Today's recipe is 75 per cent thorn, 25 per cent 'other', plus a tree in a tube every 25 yards. The 'others' in the hedge can include blackthorn, which is wonderfully intruder-proof but inclined to sucker in all directions, hazel, dogwood, guelder rose (my own favourite additive), field maple, wild rose and holly.

For many hobby farmers, working on 5–10 acres, hedges have an important part to play not just in forming the field boundaries but also defining the feel of the place. Neat privet can look too suburban beyond the garden; hawthorn may seem unsophisticated, and the mixed-species hedge is never going to be tidy; clipped beech on the other hand is always smart, a real town-and-country compromise; while yew is the finest of formal hedges but deadly to stock, which tends to limit its use to the inner garden and the family graveyard.

Taking a look around your parish, noting the hedges you like and the ones you don't, you can use a check list to keep your priorities in mind. You might give high priority to your hedges being stockproof, deterring vandals, decorative at all seasons, providing winter shelter, not too thorny (often important if clipping is to be maintained), useful to wildlife, or perhaps you need shrubs willing to grow in the shade of larger trees.

Neighbourhood hedges will give you a clear idea of what is going to thrive. Not every species, even among our native trees and shrubs, will enjoy life on chalk or in the salty blast of sea gales, on dry ridges or swampy lowlands.

There may be one overriding consideration which governs your choice, such as vandalism. In London's parks, where destructive youths are often the main problem, the last-resort plant is *Berberis stenophylla*. Its arching branches, covered with yellow flowers in May and June, and later with red berries, are fiercely thorny. Even the most determined thugs will find it easier to take another path.

This is an attractive shrub, growing to about 10ft (3m), but not one to be trifled with. It has the added virtues that pheasants eat the berries (their native food in Central Asia), no amount of cold will kill it, and it does not attract the attention of people who want to cover every surface with spray painted logos.

If there is a fashion trend in hedging, apart from the addition of some native 'others' among the quickthorn plants, it is the squeezing out of all foreign species. This is a pity because many of them have the real plus point of being ignored by browsing rabbits or deer and as a result they need less expensive protection than our own shrubs. Ribes, the mountain or flowering currant, *Rosa rugosa*, *Lonicera nitida* and even Buddleia will establish where natives are destroyed.

We have a local Buddleia hedge, self-seeded along an abandoned railway, which is a wonderful sight in high season, covered in blue flowers and multi-coloured butterflies. The scent drifts for long distances and in recent summers succeeded in drawing not only red admirals, tortoiseshells, whites, peacocks and painted ladies, but also several hummingbird hawkmoths, looking very exotic against the background of the grey North Sea. No native shrub is nearly as attractive.

This wall of Buddleia can hardly be described as a tidy hedge, towering to 15ft before it is chopped down by a tractor-mounted cutter every third or fourth year. However, it is an effective field boundary and it's there because the deer, sheep and rabbits browsed and grazed around the seedlings without touching them.

When you get around to planning the line of your hedge and ordering the plants and materials required, you begin to realise what a luxury operation this is. A post-and-wire fence 100 yards long can be planned and erected in a couple of days at almost any season. It will stand for years without maintenance. On the other hand, a hedge should be planted during a narrow window of opportunity in autumn, will take several years to establish, might not even survive at all if the summer rains fail us again, and you will probably have to put up a double post-and-wire fence anyway to protect it against stock and rabbits. The hedge will need to be trimmed, possibly every year if it extends along the side of a road.

Hedges are a beautiful and useful idea, but they are a legacy from an era when labour was cheap. At every stage they need time and money: digging the line, usually suggested as 2ft wide and at least a foot deep, notching in the two-year-old plants, keeping down the weeds with polythene or a mulch, following up with applications of glyphosate (Roundup) to give your struggling plants a chance, trimming, and maybe even laying the hedge, which is a craftsman's job.

One reference work says casually: 'In the first summer after planting, your new hedge will need watering if there is a lengthy dry spell.' That was easily written, but I wonder if the expert has ever tried to do it. Not easy.

One consolation is that your new planting does not fall within the latest law concerning anti-social hedges unless it is evergreen and more than 2 metres high (just under 7ft.).

The law is aimed quite specifically at that wonderful tree, the Leyland cypress (*Cupressocyparis x leylandii*), now routinely referred to as the 'dreaded leylandii'. This fine hybrid conifer is fast growing, windproof and adaptable, but because suburban planters fail to cut it down to size when it is used as a hedge, it has become a bogey tree.

One crusading website says that the Leyland cypress is promoted by nurserymen because 'they can grow it easily and therefore sell it cheaply'. In truth, it has to be grown from cuttings planted in mist propagation houses, not an easy technique, and this is the reason why people buy the species from nurseries rather than growing their own.

Ten thousand complaints every year pour into local authority offices because the owners of Leyland cypress hedges are too lazy or too incompetent to get out a pair of shears and trim them. Murders and suicides have been blamed on the trees' vigorous growth.

But when I look out on a bleak spring morning, with a gale whipping off the sea and across the Home Park, and see our Castlemilk Moorits and their newborn lambs sheltering in the lee of half-a-dozen Leyland cypress trees, I can only be thankful for the splendid toughness of this beautiful conifer. Even though they do not seem to fit very happily in suburbia, they are definitely the hobby farmer's friend. If ever I was take over a new place, my first move would be to get the leylandii into the ground.

And I would plant hedges.Despite the cost and the demands on labour, they are worth the effort. Nothing adds so much to the feeling that a place is being nurtured and managed. Smart fences show good management. Neat hedges show real love. And there's a great truth in the phrase 'good hedges make good neighbours.'

Meat Chickens: Just six weeks from hatching to the casserole dish

Most small-scale poultry keepers think in terms of eggs rather than meat. However, table chickens are quick and easy, actually taking less time than egg production – from day-old chick to casserole dish, the timetable can be as short as six weeks.

Even if the poultry keeper chooses one of today's slightly slower-maturing broiler hybrids advertised as being more suited to pasture rearing, the birds will be little more than eight weeks of age when taken for the freezer.

Until very recently it was difficult for the smallholder to get started with meat birds of any kind because the day-olds of the commercial breeds were available only to factory-sized units in batches of thousands. Today that picture is changing, partly because French and Belgian chicks are coming on to the market and the Continental meat birds are often being sold by the dozen rather than by the thousand. These are not only fashionable yellow-skinned chickens but also much more suited to outdoor life.

There is a wide selection of ways in which the backyard

poultry farmer can go about putting home-grown meat on the table, ranging from the fast Ross/Cobb/Hubbard broiler option, to the slow fattening of males of a dual-purpose breed, such as the Light Sussex, whose females will go on to become egg-layers.

As a starting point, 25 day-old chicks of a specialised broiler hybrid, mixed males and females as hatched, will cost about £25–£28, plus carriage. Prices for larger quantities are much cheaper – about £55 per 100. This may seem quite expensive when compared with the cost of a frozen chicken in the supermarket, but you know what your birds are eating and they are likely to be free of all the minor bugs and bacteria that seem to be present in mass-production units.

The simplest way to rear them is in a pen on the floor of an airy, concrete-floored building such as a barn or stable. A floor space about 10ft square is adequate for 25 chicks, but 15ft x 15ft is even better and makes it easier to keep them clean. A 250 watt infra-red lamp hung about 2ft above the sawdust or shavings on the floor provides the heat and the chicks are initially corralled close to the warmth by erecting a foot-high ring of corrugated cardboard inside the pen. A floor temperature of 95°F is ideal in the early stages and this is adjusted by raising or lowering the infra-red lamp.

A brood of chicks which is feeling too cold will cluster under the lamp and make a lot of noise. No matter what the thermometer says, the lamp should be lowered to increase the heat. Chicks which are too hot will scatter around the walls of their cardboard corral, gasping. By adjusting the lamp height and making sure there are no draughts, you can keep the young birds happy, gradually giving them more space and raising the lamp so that the floor temperature drops by 5 degrees a week.

Chick starter crumbs are the simple diet and provide all the birds' needs apart from clean water. The chicks will grow astonishingly and 6–7 weeks later you will have a pen full of broad-breasted, chunky birds each weighing about three and a half pounds when prepared for the oven.

This type of broiler-fryer is known in America as 'Cornish' or 'Cornish hen'. It is stocky in build, with lots of breast meat and stout legs, and descended originally from the type of fighting cock kept in the South-West of England. The Cornish miners eventually turned it into a fancy breed called Indian Game. The Americans made it into a profitable superstar – and chicken nuggets.

The Cornish hen has become the world's standard 'broiler', developed by the multinational poultry breeding companies, and billions of them are reared annually. This makes the domestic fowl the world's most abundant vertebrate. (The order of abundance: 1st. hen, 2nd. men, 3rd. rat.)

If the smallholder wants to raise freer-range poultry meat, the timetable must be extended a bit and the birds reared partly on grass. This can be done by adding an outside run to the little pen in the barn and using a slower-growing variety (say the brown 431 pasture hybrid from Sasso) and taking them through to two months of age. The breeds which mature over this slightly longer time-span tend to be stronger on their legs and they also suffer much less from the vices of feather-picking and cannibalism.

The popular French meat chicken for pasture rearing is known as the 'Master Gris', and has grey-speckled 'ermines' plumage, with yellow shanks, beak and skin. It is well adapted to fresh-air life. Although promotional material describes 'flocks of slow-maturing birds' these are also destined for the freezer at about 56–60 days. This one is

conveniently available from Meadowsweet agents throughout the UK (see appendix).

An alternative to the comparatively high-tech rearing of
meat chickens under an infra-red brooder lamp is to go back
to the peasant system and rear them under broody hens.
This may sound primitive but it is remarkably straightforward in the summer months and produces a brilliant end
product.

If you can buy or borrow the broody hens, as keepers
used to do with their pheasants, the schedule can be just as
short as with the artificial system. Buy in 25 day-old
pastured meat chicks, split them into two broods and put
them under broody hens in 10ft x 6ft movable pens in the
paddock. Let nature do the rest. Instead of worrying about
whether the young birds are warm enough, or finding their
way to the food and water, you can leave it all to the foster-
mother hen.

Keep moving the pen on to fresh grass and there are no
problems with cleaning out. Low tech is super efficient. The
chicks grow just as fast as they would indoors.

The disadvantages of pastured chicken are the familiar
ones of predators and climate. Checking the pens in the
paddock on a wet evening is tedious. Broilers and their
human attendants are more comfortable in big sheds, but
many backyarders know they get a better product by giving
their birds some access to grassland, and it is interesting that
supermarkets find they can increase prices by 200 per cent
or even as much as 400 per cent just by putting the words
'reared on free range' on the wrapper.

There is a modern mind-set, incidentally, which finds it
difficult to believe that a hen can be entrusted with the
delicate process of rearing a dozen chicks. A person who has
been taught that 240-volts-and-a-microchip make a better

job of every task can be very unwilling to hand over valuable babies to a simple broody. But she really is the best mother and because the temperature is always right, she rears the best chicken – good texture and fine flavour.

For maximum efficiency on the smallholding, it might be thought ideal to have a laying breed for eggs, a meat breed for the table, and possibly a third pen of cross-Silkies as foster-mothers for pastured broilers and replacement layers. Even better, why not have a single utility breed whose pullets could lay respectable numbers of eggs, with meaty cockerels, and provide some broody hens in season?

The search for this perfect all-around breed was pursued by poultry farmers for more than a century until the modern specialised hybrids conquered the market. However, the utility types are still out there and enjoying something of a comeback. They will never outscore the hybrids on the accountants' bottom line, but they do make sense for the hobby farmer.

The names of the utility or dual-purpose breeds are familiar even to people who have never kept any kind of poultry – the Rhode Island Red, Light Sussex, White Wyandotte, Plymouth Rock and Welsummer have stood the test of time. And there is new interest in bringing back the qualities which made them famous, improving the egg-laying abilities of the best strains beyond 200–220 eggs a year, and coupling this with the conformation and growth rate of a reasonable table bird.

There is even a modern Belgian hybridiser's 're-make' of the Light Sussex, called the Sussex Star, which is today's version of the utility, dual-purpose chicken. It is sold via Meadowsweet agents.

All of these breeds are comfortable at liberty in all weathers, will lay through November, December and

January without artificial light, and can be reasonably productive for two years. They also look good, and you can breed your own replacements.

The main skill involved lies in tracking down a good strain in the first place. Utility poultry of the heritage breeds fell on hard times after the 1950s and few breeders persevered with the recording of flock performance which is needed if the birds are not to go into decline and finish up as pretty but pointless.

And when if you decide to rear and slaughter some broilers, is home-grown pasture chicken worth the effort?

I asked a neighbour who for the past two seasons has reared about 50 French table chickens each summer. Her verdict: 'Well worth the effort, once I learned my way around our particular bottleneck, which was a lack of volunteers at plucking time. The answer was to skin the birds and store them in freezer bags. Many were being used for Italian or Indian recipes which called for skinned chicken joints in any case, so it was the easy solution.'

Milking Cow: The feelgood factor –and better dairy products than money can buy

For many centuries, the dream of the perfect rural life has been based on two acres and a milking cow. There are some pigs and hens in the picture as well, but the centrepiece is the tranquil, productive cow and her calf.

The milking cow is not the brightest of domestic animals, although endlessly curious and often quite cunning, but one of her greatest virtues is her calmness. As Michael Denny once wrote in The Observer: 'Cows can be a cure for neurosis – for angst, anomie, alienation, the blues, cafard, culture shock, future shock, identity crisis, the mean reds or whatever you've got. Everyone with a high-powered job should have one, preferably in the office, to keep them calm in moments of stress.'

And the milk is good too.

All you need to keep a cow is space and time. If you have both of those, you are probably not too stressed out in any case. However, the cow makes life even better, introducing a rhythm of feeding and milking which can be quite hypnotic.

The long squirts of milk foam into the pail, your forehead is against her flank, she feeds contentedly on her favourite titbit from the trough as you talk, encouraging her to let down the milk, and the farm cats sit in the background waiting for their share.

The routine of cow-keeping is so ancient that it seems right. The plastic quart bottle from the supermarket cabinet may be cheaper and quicker, but it will never be nearly as good for jangled nerves.

The hobby farmer's family enjoys better dairy products than money can buy and a little Jersey or dual-purpose Dexter definitely upgrades the project from backyard amusement to semi-serious farming.

The basics of husbandry for the milking cow are simple. I have seen contented cows all over the world, from Spitzbergen miners' sheds only 700 miles from the North Pole, to Masai mud huts on the Equator, and caring for them is not rocket science, but they do need someone around to milk them twice a day. You can't take a long weekend break if a cow is waiting with a full udder.

To produce milk, a cow must calve each year. A young heifer of a lightweight dairy breed such as a Jersey will be put to the bull or artificially inseminated for the first time around the age of 18 months and the gestation period is nine and a half months. When a mature milking cow is served, she will continue to produce until about 8-10 weeks before the next calf is born and then she is dried off until after the birth.

The production of milk is measured in lactations – the milking period from the birth until the cow dries off, which is usually taken as 305 days. Unless you keep more than one cow, there will be gaps in milk production. A smallholding cow has a long productive life and 12–15 years is typical.

Buying pedigree stock from the earliest stage is important. The female calves will have some value and will find a ready sale if the bloodlines are registered. The hobby farm's house cow can also help to preserve an old breed such as the rare, mahogany-coloured Gloucester, with its distinctive white stripe down the spine and famous cheese-making milk, or the tough little Shetland, suited to harsher climates and poor grazing. The best starting point is an in-calf cow with growing calf at foot.

Buying your first cow is tricky business – you need all the experience you haven't got. At least see her being milked and try to have someone with an eye for cattle take a look at the beast. Like horses and second-hand cars, there are lots of duds being offered by smooth-talking dealers. A mean, jumpy milker with a savage, pail-scattering kick is not going to calm any owner's nerves.

Another option is to avoid the milking chores completely. Crofters in the Highlands and islands with access to areas of communal grazing have traditionally kept hardy suckler cows and reared calves mainly for export to Lowland farms for fattening. For someone who has a few acres of grass and wants to keep a couple of cows but can't face the twice-a-day milking regime, this might be the way to go. Breeds like the Highland and Galloway are slow to mature – typically the cows will have their first calf at three or even four years of age – but they live out all year round with minimal feeding of concentrates and will rear their calves happily with very little attention. A few Longhorns do look good in the park.

The hobby farmer is never going to be as close to a suckler cow as he or she is to the house cow. However, there is great satisfaction in rearing good calves and they can be kept tame with some hand-feeding. Cattle are moderately intelligent – although they are not in the same league as pigs or even horses – and will respond to kind handling and titbits. The Moscow Circus used to have a troupe of cows which danced minuets, so it can be done!

The small-scale farmer can work with cattle in several ways. A house cow can be kept purely for milk – her calf sold off shortly after birth and all her production used as fresh milk, butter and cheese – or some of the milk, combined with a substitute rearing formula, can be used to pen-rear the calf. Or a couple of cows can be kept on pasture, suckling their calves and taking them right through to the age when they can be sold as weaned growing stock. Use the cheque to buy milk from the supermarket.

One of the disadvantages of the milking cow is that she is just too productive for today's average family. The dairy industry is hedged about with regulations, even more so than other aspects of agriculture, and it is impossible to sell the milk without inspections, licensing and quotas. Much of it may be wasted as a result. Even a small cow is likely to be producing more than 30lb. weight of milk every day.

Your Jersey cow is easily capable of putting four or five quarter-pound pats of beautiful butter on the table daily, besides ample fresh milk and cream. Even your best friends will begin to say 'thank you, but no,' when you offer another slab of home produce from the churn.

From Georgian times onwards, the Channel Island breeds of cattle – Guernseys and Jerseys – have been seen as the ideal 'appendages to gentlemen's parks or villas' and they were imported to many estates near the south coast ports even though they were originally seen as thin-skinned and delicate. As dairy housing improved, they became more widespread and by the early 1900s they outnumbered all breeds except Shorthorns, although both were eventually eclipsed by the high-gallon Friesian-Holstein.

The Jersey is one of the most distinctive of the world's cattle, a supreme example of the breeder's art. The Channel Islanders have kept their stock pure for nearly 250 years, allowing no live imports. The cows were tethered in the fields using a rope around the horns (they are now polled) and frequently handled, which is said to account for their pleasant temperament, and they are very easygoing with people they trust. Jersey bulls, however, are agreed to be among the most aggressive, agile and dangerous of all cattle.

'Dainty, aristocratic and almost deer-like' she may be, but the Jersey heifer is also remarkably strong. I have memories of being sledged on my belly across a wet field by a young cow which took a sudden dislike to being led on a halter. Having a 16-stone man on the other end of the rope did not slow her down at all.

Cattle do not need mesh fences like sheep and they usually escape only because they have pushed down a hedge or fence while rubbing themselves against it, which is the reason that barbed wire was invented. They are heavy and clumsy, facts to be kept in mind when building barriers – sheep tend to go over or under, cattle go straight through the middle.

Footnote: The lifestyler's cow generates a great deal of paperwork and red tape. Do not neglect to find out about registration and the bureaucratic must-do list when you buy your first milker or calf – there's quite a lot of it, from passports to insurance, including movement and health books, ear tag records, notifiable disease records, transport regulations, welfare rules and much else. All you need is a cow to keep you calm.

Paddocks: Making the best use of valuable grazing

There is no sadder sight than a sick paddock, with patches grazed bare and the whole field pockmarked with dung heaps and rampant weeds. To the less-experienced eye it looks as if only radical ploughing and re-seeding can bring the field back into productivity.

But the best tools for restoring small pastures are the grazing animals themselves and the ride-on mower from the garden. Get the timing right and your grassland can usually be made productive again without major surgery.

The point about grass is that it is the cheapest feed for livestock – less than 20 per cent of the cost of hay and concentrates. So the longer you can utilise the grazing, the better. And with some movement of the stock, it is often possible to take a crop of haylage or hay as well to tide animals over the hungriest months of the winter.

Pasture management is a local skill, with adjustments necessary if you are working on an acid, rain-sodden patch of the West Highlands or a dry, chalky upland field in Dorset. However, throughout the British Isles the climate suits the grass-clover mix which is ideal for grazing stock, and the no-growth season is mercifully short.

The most noticeably damaged paddocks are those

inhabited year-round by horses and ponies, which are very selective in their grazing habits. Even the simplest of rotations, dividing a field into perhaps four smaller areas, and grazing them in turn, can have dramatic effects.

It is possible to encourage palatable growth of grass over a long season, helping to prove that you can indeed keep one horse per acre.

The main requirements of successful horse management on grassland are to break the damaging grazing pattern by mowing – topping at a height of about 3 inches – and by introducing, even temporarily, some other type of stock, either cattle or sheep. Removing dung heaps is important, and if this is impossible, scattering them with harrows and bringing in sheep will help to even out the grazing pattern and also break the life-cycle of intestinal worms which have made the paddock 'horse sick'.

Grassland has the awkward habit of wanting to grow in one powerful surge, starting in April, growing half an inch a day at its peak in June and petering out in August, with a slight kick of re-growth in September–October. Grazing stock, on the other hand, needs a steady supply of nutrition, and management consists of keeping the grass in growing condition for as long as possible while moving the animals around to maintain even pressure of grazing.

As far as horse paddocks are concerned, topping the grass (at the time of year when your neighbours are taking a hay crop) helps to level out the rush of growth. Bringing in cattle or sheep can take off some of the excess and reduce the patchiness of the grazing pattern. A second topping with the mower in late summer will help to control perennial weeds. Cut them early enough to prevent any seeds being scattered.

On rich grassland, horses and ponies are much more likely than cows and sheep to develop problems – especially foot ailments such as laminitis. Because of the laminitis risk, there is a temptation to leave horse paddocks unfertilised. This approach, however, results in depletion of the feeding value of the remaining grass and an even more unbalanced grazing pattern. Small amounts of nitrogen fertiliser are beneficial, coupled with some mowing and rotation of animals.

Sheep are particularly useful on horse paddocks in winter because they eat swards which ponies refuse to touch and they do so without poaching the ground in the same way as cattle, which are more useful in summer. Cows eat long grass – even longer than horses – and can be the 'lead stock' in fast-growing grassland. The rotation of stock also helps to break the cycle of several parasites, because horses, cattle and sheep do not share all of the same worm infestations.

Chickens being reared for meat can be folded over a grass field in spring and there might be a slot in the rotation for a thistle-eating donkey, bramble-devouring goats or grazing geese. Just keep ringing the changes if you can.

Moving horses around a sub-divided field is quite straight-forward once they have been educated to recognise an electric fence wire or tape. Cattle may need more elaborate electric fencing or even post and wire, while sheep are best kept in place by permanent fencing of large-mesh wire.

The more decorative breeds of sheep, and particularly the short-tailed primitives such as our own adventurous Castlemilk Moorits, are likely to come to grief in electrified plastic-mesh fences. They get caught by the horns and hopelessly entangled as they try to keep up with their tiny lambs, which pop in and out through the mesh.

Movement of stock around the mini-paddocks is judged by eye. Grass in May–June grows rapidly and a patch might need to be rested for only a fortnight before it becomes so leggy that it has to be topped with the mower. In winter, a sub-division of the field may hold stock for 2–3 months, the animals supported by haylage or hay grown in one of the blocks in summer. The main reason for a move in winter may be poaching of the ground rather than the growth of the forage.

Overstocking is always a temptation, with just too many animals per acre across the whole smallholding. We have to learn to recognise land that is being undergrazed (when it needs topping) or overgrazed (when it needs a rest). As a rough rule of thumb, the grass-clover sward should be about 3–4 inches high through the season.

A paddock should be grazed before the grass gets tall and tough. The animals should be moved on before they start to damage the regrowth of the plants.

If grazing of a typical grass-clover mix is too light, the ryegrass in particular will become so rampant that the clover is shaded out, losing much of the nutrition for cattle and sheep and shortening the season. One of the difficulties with mixed grazing of horses and other stock is that grasses are much more important in horse diet than clovers – grazing seed mixtures for ponies might have 50 per cent ryegrass, 25 per cent turf grass and 25 per cent meadow grasses (timothy or creeping red fescue) without any clovers at all. However, low-growing white clover is almost essential for cattle and sheep and will be acceptable in the horse diet especially if some hay is being fed as well.

After a season or two of bad management we have in the past rescued an existing ley sowing fresh grass seed into the sward, although this should not be done at the height of the growing season when even the most patchy grass will smother new growth. Scattering seed in March and April or in the late summer and early autumn (August–September) can have brilliant results. The field should have been grazed beforehand and try to pick a wet week for the job. Run a harrow over the meadow and broadcast the seed. Sheep can be allowed into the paddock for a day or two to trample the grass seeds into the turf.

The seedling grasses will establish themselves within a month or so and light grazing can start again within five weeks. Overseeding horse grazing with perennial ryegrasses will cost about £25 per acre for seed. It's not a complicated thing to do and the results can be quite dramatic, bringing a patchy field back to life as the grazing animals utilise the entire sward, seeking the new growth among the old grasses.

Some fine-tuning of the grass-clover mix is possible. Early applications of nitrogen will favour the grasses. Later in the season, potash and phosphate will benefit the clover. Early grazing benefits the clovers, which emerge from the tightly-cropped grasses into the sunlight. Giving the paddock a break after early hard grazing will also encourage the clover.

Late application of nitrogen at about 60lb to the acre can give the grass a boost which will provide a stockpile of 'standing hay' well into the winter and sheep will graze this even in heavy snow, enjoying rank growth which they would ignore completely in the summer months.

In many parts of the British Isles, the 'magic bullet' for bad pastures was always considered to be a dose of lime. Indicators of a lack of lime are clover growing with stunted leaves, earthworms coming up out of the soil covered in small particles of soil rather than slimy, shiny and clean, and heavy colonisation by dandelions. A scattering of lime can work wonders on tired fields and is one of the standbys in situations of constant tight grazing, such as zoo paddocks, where removal of dung and regular liming keep the land remarkably fresh, year after year, without ploughing and re-seeding.

Whatever type of stock you have in your paddocks, make sure there is plenty of fresh water available at all seasons. Total freeze-ups do not usually last for long in our climate. However, if you are in a really cold spot, it might be worth experimenting with the old-fashioned idea of setting a water bucket in a nice warm bed of composted manure. One barrowload of muck at the right stage of fermentation can keep a water trough free of ice for a fortnight.

Pheasants: Rearing under broody hens, the best system

Our penned pheasants lay their first eggs in the last week of March and by April we are well into the laying season.

Game pheasants may seem rather wild and flighty in confinement, but they do lay eggs. The average production from hens in large moveable pens can reach 40 eggs per bird, and would be more if we did not turf them off our ten-acre patch and out into the wild again in late May.

For the hobby farmer who is producing birds for the family shoot, there is a lot to be said for using broody hens to hatch and rear pheasants. The sitting hen is wonderfully efficient, especially if she has been specially bred for the job, and she is immune to power outages and mechanical malfunctions. Above all, she raises the best poults, the ones that will survive after being released into the wild.

If you are a smallholder who is into the idea of being 'eco-friendly', the broody is the ideal option. If you enjoy the sight of good birds, contented and well-feathered, this is the way to go. The pheasant is the most dramatically beautiful of British birds and it is wonderful to have a few around the place even if they never get as far as the shoot coverts.

There are two starting points for the pheasant-rearing small farmer: the stock of laying pheasants and the flock of sitting hens.

The pheasants are usually caught up in the coverts in late winter (right through into March), using simple soft-mesh traps with wire-netting funnel entrances placed on pre-baited sites, and they are penned up in the ratio of one cock to five or six hens. The breeding accommodation can be either large communal pens, usually covered with netting to keep out the crows, or moveable pens on grass.

The simplest option is to put the breeding birds in a fenced, but roofless area, and make an attempt to keep the crows at bay. You will need to keep your breeding pheasants grounded by putting brails on their wings. It is even easier just to clip the flight feathers of one wing, although this makes the releasing of birds at the end of the breeding season a more extended process – they will not be able to fly until July or even August.

The persistent egg-thieving crows are quite likely to give you an ulcer, and it really is more efficient to keep your breeders under netting. You get more eggs and much less worry.

Breeding pheasants need a reasonable diet and lots of shelter. You can buy gamebird breeders' pellets, but thousands of eggs are produced each year on nothing more balanced than plain wheat. As far as shelter is concerned, little tent-style shelters of conifer branches will be sufficient, although if you have a permanent laying pen, you might consider planting some Pampas grass, whose tussocks provide perfect cover from bad weather and for laying.

Moveable pens used to house laying groups – each pen with a ground area of about 10ft x 6ft – can double up as rearing pens if you plan efficiently and they do give good control of the breeding stock. Each accommodates a cock and five hens comfortably for a short laying season – say six weeks.

It is worth setting up extra breeding groups so that you can get an early flush of eggs and move on to the rearing part of the programme. The layers are then released.

In parallel with his pheasant egg production, the farmer has to think about his foster mothers, the sitting hens. They need a much longer lead time. Silkies or silkie-cross bantams are worth their weight in gold to the bird-raiser, so you will probably have to rear your own. It is fun, but it does take a couple of seasons to build up a reliable stock which will sit at almost any time in spring and summer if you show them a nest-full of eggs.

I find that my free-range silkies come into lay in February after a short winter gap (some lay all winter), with the adult hens producing only a couple of dozen eggs before they go broody. They then continue a cycle of laying and brooding through the summer and into autumn.

The previous summer's pullets go on laying for longer. They may not sit until late April or May. After that, they follow the same pattern as the older birds. All silkie hens are fascinated by the sight of a nest with a dozen eggs and even if they are in full lay themselves they will probably find them so irresistible that they will cuddle down and start brooding.

On the subject of sitting boxes, I have tried numerous layouts. However, the problem has never been the box itself but the chore of cleaning it thoroughly after the hatch. So I now use cardboard boxes from the supermarket. These are tipped on their sides and have a wooden bar nailed across the front to provide a sill about 3 inches high to keep in the hay which forms the nest.

These boxes are readily accepted by the little silkie hens, which cover 12–14 pheasant eggs. I used to have worries about humidity and hatching problems in such dry surroundings. The hens, however, seem to adjust automatically and dead-in-shell chicks are a rarity.

The boxes are set up in small wire pens in a concrete-floored building which was formerly a cattle court. The hens are left to take exercise and feed themselves in the sawdust-bedded pens. They sit for three weeks and three days, after which the broody and her hatchlings are taken out to the rearing pen and the nest boxes go on the bonfire.

A small rearing programme to support a rough shoot might involve 10 movable pens on half an acre of grass, each pen with a broody hen and a dozen or so chicks. The aim would be to release 100-plus young pheasants and eventually put nearly half of them in the season's bag.

The design of the rearing pens is a personal choice. They do not have to be very high – just 2ft is adequate – and the easiest option for access is to hinge part of the roof, which should also incorporate a sheet of transparent roofing plastic (Corolux or something similar) to keep off the rain while letting in plenty of light. The young pheasants should be able to see the world around them, so it is important to include plenty of netting and avoid a design which has completely solid sides and ends. Poults are very difficult to settle if they have never seen anything except the walls of their pen and the sky.

To keep the ground clean, the pen should be moved once at the end of the first week, twice during the second week, three times in the third and then daily until the young pheasants are six weeks old. Regular moves are the sovereign remedy against feather picking. And if you are in a hurry, half a move, sliding one end of each pen sideways on to fresh grass, is better than none.

The diet of pheasant chicks is straightforward: game or

turkey starter crumbs for the first month, then on to growers pellets, with some kibbled grain added just before release. Grain is obviously more waterproof and can be scattered about on the release site, which is not possible with pellets. And water, of course, must always be available.

Releasing your poults, in the ideal world, is simply a case of propping up one end of the rearing pen and letting them explore the neighbouring woods and hedgerows. It is more likely, of course, that you will have to ship them some distance to a release pen in covert. The advantage of broody-reared birds is that they can to some extent be anchored by the presence of their foster parent.

Silkies make wonderful broodies. However, when they are taken to covert their big drawback becomes obvious. They cannot fly up to roost; their disintegrated feathering anchors them to the ground. This is where the cross-bred, broad-feathered hen comes into her own as she leads her brood up into the branches.

My own crossbreds are steady but also capable of flight. Even so, one three-year-old hen was caught the other day by a passing greyhound. The bird-catching dog was a delightful golden-brindle bitch who goes AWOL from home every six months or so and arrives on our doorstep. The big greyhound took the hen back as a present to her owner, the mile-long route across country marked by occasional bunches of black feathers.

I went to collect the hen the following morning and found her enjoying a luxury diet of bread and water in a cage on the kitchen table. And when she was brought back home and released to join the flock at liberty, she walked past me into the henhouse and within 10 minutes had laid an egg. Barnyard bantams are wonderfully resilient birds.

Old-Fashioned Pigs: There's real demand for high-quality pork

The domestic pig of Britain, as originally designed, was a free-ranging beast which could scrounge a living from forest and common. It was razor-backed, lop-eared, mean and hardy, feeding on acorns and roots, guided and guarded by a swineherd.

As the 'waste' was brought into cultivation, the lord of the manor lost interest in his destructive pigs, fired the swineherd and allowed his labourers to keep pigs in their back yards, with some rights giving them seasonal pannage in the forest.

As its peasant owner was also gradually squeezed off the common land, the pig moved into a sty, where it was fed on garbage and was the main source of desperately-needed fat. Crossing with imported Chinese snub-nosed pigs in the 1700s helped in this transformation, but also made the resulting animal much less hardy.

Eventually, mass-production of pork and bacon moved the lardy pig from the smallholder's back yard to the modern high-tech unit. The slimline landrace/large white breed was triumphant. It looked like the end of the story.

But nobody had consulted to the customer about all these changes. And the modern buyer of bacon, ham and pork was not always delighted with the product. There was an underlying demand for something better than packaged pork which looked like compressed newspaper pulp and tasted very similar. This market for good pork from traditional breeds is one of the few areas in which today's hobby farmer is being offered money by active buyers for something he can produce at a profit – and he can help to conserve rare breeds while doing it.

The rare breeds descend from a not very well differentiated all-England pig which existed in the early nineteenth century. Some parts of the country liked a white pig, some liked a curly coat, some preferred a spotted pig and others thought the best ham undoubtedly came from a lop-eared black pig with a white belt round its middle. But all of these pigs were very similar in their design, which was based on crosses between the razorback hog of medieval times and the prick-eared, wonderfully fat little pigs from China.

As some British breeds – notably the large white – became dominant, local variations on the basic pig began to slip into extinction. We said goodbye to the Ulster White, the Dorset Gold Tip, Yorkshire Blue, Cumberland and the Lincolnshire Curly Coated. We were on the point of losing the Tamworth, Middle White and several others.

But then there came a bit of a resurgence as the backlash against intensive farming methods began to bring a small premium for free-range pork. There was new enthusiasm for breeds such as the Gloucestershire Old Spots, whose breed society was formed as late as 1913, nearly half a century after similar societies had been established for most breeds of cattle and sheep. In its heyday, it had been promoted as the 'orchard pig' and the 'cottage pig', just the tags that made it once again fashionable as the 21st century dawned.

Today more than half a dozen rare breeds are hoping for a revival in their fortunes. The Gloucestershire Old Spots is one of the varieties under the wing of the Rare Breeds Survival Trust, along with the British Saddleback, Berkshire, Large Black, British Lop, Middle White and Tamworth. These each have between 70 and 110 breeders.

Pig breeds are much faster at climbing out of a near-extinction trough than other farm livestock because of their rapid breeding cycle and large litters. Where an owner of a group of three breeding females of a rare sheep breed might be happy to have three lambs at the end of the year, the owner of three sows could, with average luck, have more than 60 piglets. They go from bust to boom very quickly.

There is, incidentally, a curious downside to the fast breeding abilities of our pigs. Not only can you save a rare breed more easily, but you can re-create one, which upsets some people. A case in point is the story of the Oxford Sandy and Black pig, which was recently described as 'a running sore' in the history of the Rare Breeds Survival Trust. It is still not accepted as a rare breed by the RBST. However, it is an extremely stylish and colourful pig, a natural browser and forager according to its fans, so even though it is not listed as an official rarity you might find it interesting, especially if you live in Oxfordshire and you wanted to arouse the enthusiasm of local butchers for Oxford's own ham, made from the 'plum pudding pig".

An alternative scenario for the creative marketeer is to manufacture your own local pig from scratch, which is exactly what livestock breeders were doing a century ago with everything from swine to poultry. You take a variety you like – say the Oxford Sandy and Black – make it your own by changing an important feature (perhaps the long snout could become snub-nosed by using a Middle White boar) and you can say, without blushing, that the 'exact origins of the breed are lost in antiquity.' In a few seasons you have your 'Hereford Cyderbucket pig', 'Ayrshire Milksweet pig' or whatever, then promote it to the local ham market.

This would be particularly appropriate in a county like Ayrshire, which has its own famous bacon cure but no breed of its own to complete the picture. Wiltshire hams, York hams and Cumberland sausages are surely crying out to be linked to local breeds of pig.

The meat produced by these old breeds, whatever their names, is agreed to be very good indeed. Derek Cooper, writing in BBC Good Food magazine, said of Gloucestershire Old Spots pork: 'Everyone was impressed by the quality and flavour of the pork and its crisp crackling. It was a rare treat...'

John Bailey, of Johnsons of Yoxall, Staffordshire, recently voted the RBST accredited butcher of the year, said: 'The stars are Gloucestershire Old Spots, Tamworths and Saddlebacks, which all produce great crackling."

Antony Worral Thompson, known to many as the resident chef on BBC's *Food and Drink* programme and a regular on Ready, Steady, Cook, is a great believer in rare-breed pork and now keeps his own herd of Middle Whites to produce top quality meat.

The attractions of pig-keeping for the smallholder often lie in the sty rather than in the paddock. A few pigs can turn a plantation or orchard into a horrible quagmire in a few days. As one friend says: 'We would like to use pigs to do a bit of ploughing, but we are on silty clay and a pig left to its own devices will take the structure out of the soil in less than a week...leaving it resembling concrete."

Looking after pigs on open range is not much fun in bad

weather and I'm not sure it's always much fun for the pigs, either. Our place sits 300ft up, looking out to the North Sea, and I have recent memories of seeing a neighbour throwing dozens of blue, frozen piglets from his arks into a trailer on a February morning. He has since given up – gone to work as a bus driver – and the new owner of the farm keeps his pigs inside. He says 'free range' killed most of his farming forebears, who died of bronchitis and rheumatism, and he would rather keep some walls between himself and the winds from Siberia!

If you are taking the progeny of three sows on to marketable weights, you might need to handle 12 tons of commercial feed every year, which is hard work for the hobby farmer, not a minor chore you can hand over to an obliging neighbour while you pop off for a week in Biarritz.

And you can no longer use the pig to do what it does best, which is recycle leftovers from kitchen. That is now strictly illegal, seen as the source of outbreaks of swine fever and foot-and-mouth disease. Some of us, brought up in more frugal times, see it as insanely wicked to flush food down the drain while the pig is out there in the yard waiting to be fed, but the law is firm on the subject.

If you have the energy and the time, the pig is in demand. The Rare Breeds Survival Trust (see appendix) operates a meat marketing scheme and lists approved finishing units around the country, nearly 20 of which are currently looking for pigs. And the RBST premium adds nearly 50 per cent to the value of pork pigs when compared with the ordinary commercial market.

Poplars: Fast, vigorous – and they look good

There is something wonderfully satisfying about growing hybrid poplars. In an age of instant gratification, the modern clones of this beautiful broadleaf are rewardingly quick, dramatic and almost foolproof.

Compared with most projects around the farm and garden, growing poplars is uncomplicated. One fine day, between early November and May Day, you cut a stem of poplar sapling, about 2–3 feet long and probably no thicker than your finger, poke half its length into the ground – and stand by for the action.

In late spring, the unpromising twig bursts into life and sends out pale green leaves, rather leathery in texture with whitish backs, which flutter in the breeze but seem to survive every gale. As the months pass, each new relay of foliage is larger than the last, until in September your *Populus x interamericana* stands 6–8ft high, topped by a large crown of teaplate-sized leaves.

These hybrid cottonwood-poplars are among the fastest-growing plants known. Experiments in the north-western United States demonstrated that they are ten times more productive than such mainstay species of the pulp industry as the Douglas fir. And they are not only vigorous, they look good too. You have to be a very committed 'native species' devotee not to agree that a mature hybrid poplar is a fine sight in the landscape.

So fine that one admirer of my young trees suggested that our plans for an apple orchard were a waste of time – 'Who's going to pick all those apples?' – and the paddock would look better with some widely-spaced poplars and grazing sheep.

Most poplars are grown for veneers, firewood and pulp. The timber is white and sweet, without resins or splinters, which makes it ideal for food packaging. However, around the hobby farm the main function of the new poplars is likely to be rapidly-grown summer shelter, and on larger acreages they not only help to hold pheasants but provide a high horizon to get the birds on the wing.

Forest research in Britain has sometimes shown that hybrid poplars can suffer setbacks when affected by drought and squirrels or smothered by weeds in the early stages, but on a smaller scale they are resilient. I have had a failure rate of less than 1 per cent with cuttings poked into weedkilled spots, and my only regret is that I didn't plant more of them 15 years earlier. They have been free of disease, and the most recent hybrids have shown themselves resistant to canker and the various leaf rusts which decimated early experimental plots of poplar.

Our 10 acres have rich soil but cruel exposure to easterly gales and our experiments with a wide range of trees, from Australian gums to Norway maples, have had mixed results. The species have included some very satisfactory *Populus robusta* and also pure *Populus trichocarpa*, the western cottonwood, with the less rewarding *P.serotina* and various tall, narrow poplar clones which look rather awkward in our landscape. But by far the most successful has been a cross between *P.trichocarpa* and *P.deltoides*, listed as *Populus interamericana* Beaupre, a generously foliaged variety which we bought originally as a parcel of mail-order setts from a small ad in The Field.

Traditionalists will cry out in anguish, but undoubtedly the best shelter-belt combination here has consisted of Beaupre poplars sandwiching a core of the much-maligned Leyland cypress. As the poplars rush skyward, alternate stems are cut to encourage new growth from the base and to keep the whole thing solid. Rugosa roses, *Eleagnus ebbingei* and some snowberry along the fence-line provide windproofing at ground level.

Poplars grow best on agricultural land rather than uplands or marshes, and there are ambitious plans to develop a poplar-based wood products industry in South-West England, in an area from Herefordshire to Devon. For hobby farmers who think subsidised trees are the way to go – and Britain imports £1million-worth of timber every hour, every day – there are various joint venture schemes, such as those operated by the Poplar Tree Company, Lower Lulham, Madley, Herefordshire HR2 9JJ (01981 250 253).

For smaller projects, poplar cuttings (£74 per 100) are sold by Bowhayes Farm, Venn Ottery, Ottery St.Mary, Devon EX11 1RX (01404 812 229), who also have osier willows, *Salix viminalis hybridus*, at the same price.

The willows are more suitable for boggy sites, and according to research at John Moores University, Liverpool, the hybrid osiers 'have considerable potential for cleaning up contaminated land', speeding the breakdown of pollutants such as oil residues. They produce more than 15 tonnes of dry matter per hectare each year while doing it. The wood is used for generating electricity – and the thinnings make splendid goat fodder.

New Potato: Having fun with a most forgiving tuber

When dandelions come into bloom, it's time to get your seed potatoes into the ground. That's the tradition, and like many traditions it may not be 100 per cent reliable but it does give you a bit of encouragement to get started – the earliest potatoes are undoubtedly the best.

If you are a real goody-two-shoes of a gardener, you've probably had your seed potatoes sprouting for weeks by then, all stacked up in their tomato boxes, carefully padded with newspaper and encouraged to start into growth by a steady temperature in the low 40s Fahrenheit (say 7–8°C) and ample light.

For most of us, though, this is catch-up time. The spring is springing and many of the seedsmen's catalogues have a last-orders date in early March. Luckily, the potato is a most forgiving tuber and whether you are planting first-early varieties in March or maincrops as late as July, you are unlikely to have a total failure. And even if you have only a plot which you never got around to digging during the winter, or perhaps a few plastic pots lying around in the greenhouse, you can still look forward to new potatoes in season.

For the small-scale farmer who takes the potato crop seriously, a plot measuring about 10 paces by 5 paces will produce nearly 500lb to store through the winter. For the hobbyist who wants some fun, a few 12in. pots in a frost-free greenhouse will have baby potatoes on the menu only 8–10 weeks after planting. Use tall pots, put your seed potato (first-early varieties have names like Swift, Premiere and Rocket) on a couple of inches of compost in the bottom of the well-drained pot and top up the compost as the plant grows so that tubers can form up the stems. Delicious fast food.

There are quick-growing earlies, which go straight from garden to pan, and there are slow-growing maincrops which keep in store for months. There are floury potatoes and waxy ones, big varieties and tiny fingerlings, and a surprising rainbow of colours ranging from purple, blue and dark red through buff and golden yellow to white and pink. They've got wonderful names too. If you fancy growing a potato called Beauty of Bute, or La Negress du Poitou, it's in a catalogue somewhere. Shopping around Europe, there are about 4000 cultivars altogether (see www.europotato.org).

The 'industrial' potatoes which fill our supermarket bins and freezers, such as Maris Peer and Maris Piper, are dominant because they are reliable producers of heavy crops. Few of them claim to have superior flavour or texture, although on the salad counter, sales of tasty varieties such as yellow Charlotte are increasing, an encouraging trend. More than 150 varieties are available as commercial seed in this country, but nearly 70 per cent of earlies and 50 per cent of maincrops grown for the market are from just six varieties, so if you feel like branching out and growing something new, there is plenty of scope.

It's worth noting that Britain's farm yields of potatoes have doubled per acre over the past 40 years and the familiar names such as Majestic, King Edward and Kerr's Pink have been replaced by new ones. Only four of the top ten varieties of 1984 were still in the top ten in 2000. This is partly due to the demands of the supermarkets, which sell 80 per cent of our potatoes. They not only want a low-cost potato but one that looks good. The skin must be blemish-free.

The basics of growing the maincrop potato are truly simple. Millions of people around the temperate world grow them because they are a great food and they are easy. Plant some egg-sized potatoes in the ground, about six inches deep and a foot apart, with the 'eye' sprouts at the top. A pound of seed potatoes plants a 10ft row. Cover them up and when their leaves begin to appear, hoe up the soil along the row so that each plant is almost covered, repeating this a couple of times until the ridge is about 6–8 inches high. The developing tubers grow *above* the seed potato and this ridging protects them. Come back in five months, when the foliage dies, and lift your crop. That simple.

Even easier is the 'no dig' method pioneered by the Henry Doubleday Research Association, whose gardeners put chitted (that is, pre-sprouted) seed potatoes on well-fertilised ground in April and cover them with six inches of hay or old straw. Just lay the earlies on the soil about a foot apart – you can push them into the soil with the help of a trowel, but it isn't vital – and wait for the shoots to appear through the mulch. Keep adding hay if there is a serious risk of frost.

When you begin to mow your lawns, add the grass clippings to the mulch. As each layer of clippings dries, add some more until your mulch is about a foot thick, with the potato plants emerging through it. The clippings keep out light, suppress weeds and hold in moisture.

The mulch does not seem to encourage slugs – or at least no more than any other method – and it certainly hampers the activities of eelworms.

When you think your potatoes are ready, which is indicated by flowering in many early varieties, turn over a fork-full of the mulch and take a look. If they are still too small and unripe – they taste like blobs of pond-water if you harvest them too soon – just roll the hay back into place and give them another day or two. You don't have to destroy the entire plant to take a few tubers, so pick and come again. Full details of the system are on the website at www.hdra.org.uk.

The HDRA used to recommend the use of black polythene sheeting rather than mulch. However, they are now against the use of plastic on environmental grounds. Although not so protective against frost, it is a wonderfully easy way to produce early potatoes. Trowel out shallow holes for your seed potatoes (Swift is the recommended variety because the plants are very compact) and put them into the soil with the 'rose' end – the one with most buds – uppermost and just below the level of the soil surface. As with other methods, spacing of about a foot apart is suitable for early varieties.

Because protection against frost is not so effective as it is with mulching or the traditional ridging, timing is important. No matter what your local dandelions are doing, try late March in the South and April in the North. Nothing will destroy your faith in global warming more completely than the sight of a row of frost blackened potato plants.

Once your seed potatoes are in the ground, cover them with a strip of 20 inch wide, 150 gauge black polythene. Bury the edges and weight the sheet down with bricks. If the wind gets under the edge it will take the whole covering away in a few minutes and wrap it around a distant fence.

The growing shoots will force up the plastic sheet. Using a sharp blade, cut a slit or cross about 2 inches long for each plant and help the leaves to emerge into the sunlight. Make sure you do not cut too big a hole; light may get in and green the tubers. If you do slash too big an aperture, pile up some peat around the stems to keep out the light. If you have been too optimistic and find that frost is still forecast, put horticultural fleece on top of the plants overnight.

When flowers appear, you can peel back some of the sheeting and gather your potatoes for the kitchen. Most of them will be lying on the surface of the ground and need nothing more than a quick rinse under the tap.

For the small-scale farmer, there's a growing market for gourmet potatoes. Chefs are enthusiastic, looking especially for exotic fingerlings to accompany game. Several fashionable dishes, such as tiny mountain potatoes sauteed with leeks in olive oil, are North American but use European varieties, including widely-available Ratte, Pink Fir Apple and Anya.

Suppliers offer sample packs of their unusual cultivars, so whether you are trying to create a red, white and blue salad or the perfect plate of chips, there's a potato for you, just waiting to be planted next spring.

Silkie Hens: The best of all foster mothers

If you've got a few silkie bantams in your poultry yard, and a milking goat, you can rear almost anything.

I think my father would have added a third vital aid, a warm oven. He was a great believer in popping blue-chilled orphans into the oven beside the kitchen fire. They went in apparently dead in many cases – lambs, piglets, Californian rabbits, fancy mice, ducklings, canaries – and came out warm and hungry just a few minutes later. In the early days of life, warmth beats antibiotics every time.

The silkie is in the warmth business. She is the best of foster mothers for all sorts of birds. I have seen buzzards and falcons, quail and peacocks, all peering out happily from the furry feathering of a clucking silkie. In extremis, she will even attempt to brood a kitten or tiny puppy.

The virtues of the silkie were brought to mind when I was reading the small ads in our local paper. This ancient breed of poultry – first described by Marco Polo, who thought it was a cross between a hen and a cat – seems to be having a bit of a boom according to the 'Poultry and Game for Sale' column, with many more silkies on offer than laying pullets or fattening turkeys. And this exotic little pet deserves her star status; she really is invaluable for game rearing, for producing eggs and, in a curious way, on the table.

The odd fact about the breed as a roast chicken is that even white silkies (they come in several colours including white, black, blue and gold) are all black at heart. Their skins are deep violet black and their bones are covered in black membrane. On the table, the flesh appears as white veined with black, like expensive marble.

This appearance has not been to everyone's taste, as it were, with the great poultry writer Lewis Wright noting: '...the fowl, though really excellent eating, is rather repellent to ordinary notions upon a dish.'

Today it is in demand by gourmets, who call it 'Chinese black chicken' and pay premium prices for plump silkies to go into Oriental chicken pie – one of the reasons for the breed's popularity in some markets – although they are not very large, with males weighing about 3lb and hens not much over 2lb.

The good points of the silkie for the smallholder are her docile nature and poor flying ability. The disintegrated webbing of the feathers prevents silkies from flying strongly enough to get over a 3ft fence, which makes it easy to keep them within bounds, although it has to be admitted that a low fence is not going to keep out chicken-hunting foxes.

Our birds are kept in the walled garden, an ideal setting, and they can't fly up into the apple trees and then on to the top of the wall, as almost every other form of poultry and game tries to do. They stay where they are put and look very decorative, with their mulberry-coloured faces, dark eyes, neat topknots and furry legs.

Silkies originated in the East, and in this country they have always been popular with followers of the exotic, who seem to have included a lot of titled ladies and parsons. The Rev. R.S. Woodgate, of Pembury Hall, Kent, writing just over a century ago, said: 'These attractive fowls are good layers of small, cream-coloured eggs all year round. The hens will lay with cackling delight through the deepest snowstorm, and seem hard as bricks. They are very bright and contented, and a trio or two on a lawn, when they can have a small roosting-house in a neighbouring shrubbery, are not only most beautiful and attractive, but they decidedly pay their way.'

The pullets lay up to a couple of dozen eggs before settling down to broodiness, but the mature hens usually settle down after as few as 8–10 eggs and will sit, hatch, rear, and sit again, if that's what their owner wants. A spring-time advertisement offering 'Broodies for sale, buyer collects' has attracted more responses than anything else we have ever sold off our 10 acres. And they made higher prices than any layer or broiler, at up to £10 apiece.

Another plus point about silkies is that the males are just as good-natured as the hens, fussing about happily without making much noise or ever getting into a fight.

When I worked at the Eley Game Advisory Station (now the Game Conservancy) at Burgate Manor, Hampshire, we used hens for hatching all sorts of oddball experimental broods, including prairie chickens and capercaillie. The sitters were largely silkie in origin, but persistent problems with scaly-leg mite infestations, to which the silkies are very prone (your vet can give you the cure – Ivormectin) had led to crossing with Orpington bantams to produce a small, teapot shaped hen, broad-feathered, clean-legged and top-knotted, which would sit forever.

Hatching eggs from around the country could be picked up at any stage of incubation and brought back to Burgate under hens sitting in cardboard boxes in the back of the car. They never, ever stood up or made a fuss, and they did not have scurfy legs.

If we culled the cockerels, they had black skin and bone, just like the silkies from which they had been bred several generations earlier. Crosses of this type were marketed about 1900 as 'Negro fowls', although the breed never caught on.

No doubt the description itself would be culled today, but it made the point, and there could be a demand for similar birds if an enterprising hobby farmer decided to create something different for the poultry market. The only problem might lie in finding a politically correct name for the pie-makers' new black bird.

Garden Poultry: The backyard egg producer is back in fashion

Garden poultry have come back into fashion. The cackling of laying hens is being heard again in suburban backyards and the orchards of country houses.

This is boom time for home-grown fresh eggs thanks to modern feeds and greatly improved breeds. Keeping a few hens has always been a fairly simple business – the domestic fowl is the world's commonest bird because it is so easy to breed and rear – but today it is even more straightforward for the backyard hobbyist.

Just after World War II, when garden poultry hit their previous peak, the laying hens were producing about 150–200 eggs a year and their feed mixture was compounded in the kitchen, using household scraps, grain, boiled vegetables and potatoes and much else besides. Today an average of 300-plus eggs is being achieved on free range and the balanced pellet diet comes out of a bag.

The sudden surge in the popularity of garden poultry, which is almost reaching the scale of the Poultry Mania of mid-Victorian times, is prompted by our demand for healthier food. The availability of purpose-bred varieties which thrive under semi-liberty conditions makes it possible today to produce good eggs under backyard conditions with very little difficulty.

Under intensive management, when kept in laying cages with artificial light regimes, some breeds are well-known for producing astonishing numbers of eggs. According to the Guinness Book of Records, the egg-laying record is held by a White Leghorn which laid 371 eggs in 364 days in an official test at the University of Missouri in 1979.

Figures like that seemed to be fantasy for the average garden poultry keeper until geneticists turned their attention to improving free-range hens, making their hybrids hardier, better feathered, calmer in temperament and slightly more adaptable in their dietary needs. The result is a group of modern brown-egg breeds well-suited to the garden pen such as the popular Bovans Nera and Black Rocks, based on the North American Rhode Island Red and Barred Plymouth Rock (illustrated), and also Warrens (a development of the Rhode Island Red), Speckledy Hens (Marans hybrid, dark brown egg), ISA Browns, Meadowsweets, and the white-egg free-range hybrids based on the White Leghorn.

The highest recorded annual flock average per bird is 315 eggs in 52 weeks from a Welsh farm's stock of 5,997 free-range ISA Brown layers.

So 10 hens could, in theory, give you nearly 3000 eggs in a year.

If you keep a dual-purpose, meat-and-eggs breed, a realistic total over the laying cycle of 12–14 months is more likely to be 2000–2500. Each hen will be eating about 5lb of food for every dozen brown eggs on the table. White-eggers will be laying slightly more and eating slightly less, but they are not table birds.

Pullets of a laying breed will cost about £6–£8 each if you buy them at 18–20 weeks of age, just as they are beginning to lay their first eggs, and this is the starting point recommended for novices.

British backyard poultry keepers like brown eggs. No one can actually tell the difference between a white egg and a brown one once you have thrown away the shell, but they are seen as more attractive, more 'natural'. These brown-shelled eggs are traditionally produced by heavy, meaty breeds with red ear-lobes, and the white-shelled eggs by lightweight birds with white ear-lobes, although this distinction is blurred by recent hybridisation programmes. Hybridisers have had a tremendous struggle to create a really efficient brown-shell breed for the UK market. Americans prefer a white egg, so this has been a British problem.

A little group of hens can be kept in permanent or movable accommodation, everywhere from city garden to open farm fields. The permanent hen-house is often an adapted out-house, kennel or stable and most buildings are suitable if you can keep out the rain and rats and let in some air. A flock of hens will forgive almost anything except damp.

There are many ways in which buildings and yards can be adapted for laying hens. A covered yard or airy cart-shed fronted with wire-netting needs almost no fittings at all if there is plenty of dry straw on the floor and this is the ideal set-up for rearing your own broilers, which can produce large quantities of chicken for the freezer in just a couple of months from day-old.

A permanent henhouse for laying birds in the garden can have two or three runs built alongside which can be rested alternately to allow the grass to freshen up. This is one of the most successful small-scale systems. A 6ft fence of wire netting – or even string netting – will keep most brown-egg layers within bounds. Lighter breeds may need to have their wings clipped, the primary feathers cut on one wing to unbalance the flier, or have the netting extended to cover the top of the pen.

The movable henhouse can be much smaller than a permanent building if it is lifted on to fresh ground occasionally and it usually has a run attached. Your hens' accommodation is going to last for many years, so make sure it is well ventilated and that it is easy to collect eggs, straightforward to feed the birds in wet weather and simple to clean out.

Buying the biggest henhouse that fits into your budget makes good sense. 'The sizes given in most price lists are not large enough for the numbers usually stated with them – thus a house 4 feet square is often given as 'suitable for 12 fowls.' It is nothing of the sort: more than half that should not be placed in it.' The great poultry breeder Lewis Wright wrote those words in 1899, and the price lists are as optimistic as ever about the number of birds that can be squeezed into a tiny shed.

As a rough guide, if there is no extensive run or yard, 5 sq. ft. of floor space is needed for every bird – six birds will need a shed no smaller than 6ft x 5ft., with plenty of openings for light and air. Under more spacious systems, with an open run alongside the shed, the square footage of the house itself can be reduced to 3 sq. ft. per bird.

Hens greatly enjoy sunshine and dust baths. The best way to extend their house is to put up a sheet of clear corrugated plastic roofing (6ft. x 2ft.6in.) on a frame about 2ft above ground level to form a sunbathing area, although you have to make sure that agile birds do not use this as a spring-board to escape into the wider garden. Even a sheet of corrugated iron over a patch of dry earth will be appreciated.

Foods and feeding are simplicity itself. Laying hens eat compounded layers' pellets, fed ad lib in hoppers, with grit and plenty of fresh water. Half a dozen birds will eat about one bag (half hundredweight or 25kg) in a month. Extras such as cabbages will keep the birds amused and help egg colour, and titbits keep them tame, but they are not essential. On genuine free-range, with birds at liberty around the farm, the mainstay of the diet can be wheat.

Beginners often worry about poultry diseases. In practice, these are rarely a problem for the small flock. At Kintaline Poultry Centre in Argyll, for example, they say that their Black Rock layers have 'great disease resistance – we have rarely needed to do anything.' And some birds go on laying for up to 8 years.

There are more likely to be problems with noise, if you have near neighbours who are happier with rock music than Black Rocks, and with predators. The noise is reduced if hens are kept without a crowing cock bird – unnecessary for the production of table eggs – and most foxes are deterred by an electric fence wire. Cats are unlikely to attack poultry of large breeds.

The standard backyarder's bird used to be listed in advertisements as 'RIR X LS POL", which translates as 'Rhode Island Red cross Light Sussex point-of-lay' pullets. These are not only dual-purpose birds producing eggs and meat but also easily sexed at day-old: the females of this cross are dark like the male Rhode Island while the unwanted cockerel chicks have the pale down of the female parent.

This type of visual sexing at hatching was extended by breeders using barred-plumage types and this led in the 1930s to the autosexing breeds. Cambridge geneticists did away with the need for endless crossing between existing breeds when they created these exciting pure-breds whose chicks can be instantly sexed at day-old by their down colour. They are currently having a major resurgence of popularity. The Cream Legbar, which lays blue or blue-green eggs, the brown-egg Welbar and other autosexing breeds are in such demand that fertile eggs for hatching sell for £2 each at poultry auctions.

Poultry fads are famous for being boom-and-bust affairs, but while the going is good, a dozen hens of an autosexing breed – each bird laying say 200 eggs at £2 each – sounds like a bright idea. The investment is not enormous and the returns beat anything on the stock market, while they last. (The financial warning says simply: fads can go down as well as up.)

There are also high prices at auction for several ornamental breeds, such as White-crested Black Polish, which are beetle-green black with white pompom headgear, and curiosities including the fluffy-plumaged Silkies in black, gold or white, so you might soon find yourself caught up in all the fascinating aspects of breeding poultry, with bantams in the shrubbery and gamecocks in the yard.

Japanese Quail: Most productive of domestic birds

There is a lot of good news and two bits of bad news about the Japanese quail. The good news is that this little gamebird is the most super-productive of all domestic birds, laying from the age of just six or seven weeks, with table birds ready for market at 50 days. The bad news for the shooting man is that the quail is a complete failure as a wild gamebird, and the problem for the hobby farmer is that the British market for quail meat and eggs is described as 'saturated'.

For many years, shooting men in search of a quick fix for declining game numbers have spotted the potential of these coturnix quail. They have noted that before the quail hen is nine weeks old, some of her chicks will already be hatching. By the time she reaches 18 weeks, her grandchildren are themselves breaking out of their shells.

The theory is exciting: you could have three generations of birds in one summer and be up to your knees in quail before you could whistle wet-my-lips (which the wild quail do all the time).

Sadly, it just doesn't work out like that. The domestic quail is no more suited to life in the wild than a battery hen, with the added difficulty that its ancestors were highly migratory, so your hordes of little mini-partridges would be thinking of heading south by the middle of August.

Some genetic engineer is no doubt working somewhere on a mega-prolific, home-loving gamebird created by hybridising quail and partridges. But until that wonderbird comes along, the domestic Japanese quail should be treated as the brilliant result of traditional breeding skills, unequalled in its ability to produce meat and eggs for the table to a short timetable and thriving under intensive management.

These little birds can be kept in a garden aviary, fed on seed and treated like mini-pheasants, but even under these protected conditions they will rarely incubate their own eggs and the laying season will be confined to May–August. It is not their preferred habitat. If you take them indoors, give them a diet of starter crumbs with almost constant light and a friendly temperature, they become different creatures altogether, enthusiastically breeding and laying all year round.

At six weeks of age both sexes are similar in size – about 7 inches long – and also in weight. Beyond that age the males become sexually active and gain little weight, while the hens continue to grow to their maximum of around 6oz, about 20 per cent heavier than the cocks.

A breeding unit of these quail can give the hobby farmer a steady supply of both eggs and meat. These are familiar items in the delicatessen trade and it's a great pleasure to produce your own from a few pens in the barn. Quail have the added 'plus' that they do actually turn up on the home menu, unlike the 'meat' rabbits and 'meat' geese which acquire names and become farmyard pets, never getting as far as the dinner table.

Japanese quail come in several 'colourways' – Manchurian Golden, American Range, Fawn, Tuxedo and Dark-eyed White. The Range variety is rather similar in colour to a melanistic hen pheasant and the neatly-named Tuxedo is similar but with a white face and underparts, the Golden is paler than the normal form and the Fawn paler still. All these colours are equally prolific, and in addition there are also selected laying strains and the Jumbo or Giant quail, which may be twice as heavy at 6 weeks.

The names given to the domestic quail are extraordinarily confusing. There are half a dozen species in the genus Coturnix, ranging across most of the Old World and Australia. However, the 'coturnix' on its own has become an alternative name for the domestic form of the Japanese quail (*Coturnix japonica*), a bird which has nearly 100 other labels – Pharaoh quail, Holy Land quail and Bible quail among them – and including some which are correctly applied to other species entirely, such as King quail and Button quail.

One of the reasons for the mishmash of names is that the Japanese quail has never been an exhibition bird: it arrived in Britain too late to be part of the Victorian poultry mania which put standardised breeds on the show bench (as well as improving their meat and laying qualities). One minor side effect of the craze for showing breeds was an agreement on what each variety was called, a bonus which did not extend to the Japanese-Pharoah-Nile-King's-German-or-whatever quail. Even today no one is quite sure where the domestic quail fits, with stock for sale in the classified ads just as likely to turn up under the heading 'Poultry and Game' as under 'Cage and Aviary Birds'.

Even when kept under ideal conditions out of doors, Japanese quail are unreliable parents. They lay all summer, but rarely sit. I have hatched and reared them under very small silkie bantams in outdoor pens – the result was very beautifully feathered birds – but usually they are hatched in incubators and reared in brooders. The chicks, which hatch after 16–17 days, are tiny – four to the ounce – and full of energy. They are best kept on paper floors and fed on game or turkey starter crumbs (about 28 per cent protein), which also form the best diet for adults.

At quite an early age, Japanese quail are easy to sex. Apart from the distinctive brick-red throat of the male, he can be distinguished when in breeding condition by his secretion from a swollen vent gland of a white discharge from the kidneys which looks like shaving-cream foam, a unique feature of the species.

For the production of hatching eggs, a cock bird will usually be housed with two or three hens in a wire-floored cage about 20–24in. square and 12in. high. Wire mesh of 7mm (0.28in.) is recommended for the flooring. Quail can also be kept in groups on deep litter under a 60 watt bulb, lit for 16–18 hours (or even 24 hours) a day, housing one male with every three females and allowing 20 square inches per bird, although many eggs may be buried and lost when using this 'easycare' system.

One warning about housing quail on the floor: there is something about the smell or sound of these birds which seems to attract rats with an almost Pied Piper intensity – they arrive from miles around. Make certain that the building housing your quail is 100 per cent rat proof.

The migratory quail of Europe and Asia (*Coturnix coturnix*) is still around as a wild bird, although much rarer than it once was. I have heard its 'QUICK, quick-ic' (wet-my-lips) call in many places in this country, from Salisbury Plain to Kincardineshire clifftops, but it's an uncommon event, always worth noting in the diary.

Once it was a familiar sound of the countryside, and quail were so abundant on migration that they could be scooped up with nets in the streets of Egyptian villages or decoyed into traps by the bucketful on Italian hillsides. The average harvest on the little island of Capri, in the Bay of Naples, exceeded 160,000 birds a year, and ships were said to have foundered in the Mediterranean under the weight of quails when they dropped exhausted on deck.

The quail is the only bird whose numbers in Europe were ever described in terms equivalent to the vast flocks of passenger pigeons in North America. The migrants in autumn, flying south to Africa, were numbered in millions, and Asian birds wintering on the plains of India were described as 'abounding in such degree that shooting them is mere slaughter. A tolerably good shot will bag 50-60 brace in about three hours.' And that was in the days of muzzle-loading shotguns.

This little gamebird was never super-abundant in the British Isles, although quite familiar on the summer table in medieval times. It declined in the seventeenth and eighteenth centuries, and crashed in the mid-1800s. Every now and again there are good quail years, when warm southerly winds in spring encourage migrants to overshoot their Continental strongholds, but today it is an uncommon bird with us, even on its preferred habitat of dry chalk uplands. Maybe global warming will bring it back, although our current series of warmer summers also seems to bring torrential rain, which does not suit quail chicks at all.

Although illustrations in bird books depict it as a miniature partridge, the quail looks very different on the wing. The body is cigar-like rather than dumpy and the wings are long, narrow and noticeably sickle-shaped. As might be expected from a long-distance migrant, it can fly far and fast, but when flushed on its breeding grounds it usually flutters down, with a rather lark-like descent, within 100 yards of the place where it rose.

Table Rabbits: Breed bunnies and feed the world

The French are very fond of rabbit. The average Frenchman eats nearly 10lb. (more than 4kg) of rabbit meat each year and every smallholder has a few hutches in the back yard.

Blanquettes and casseroles of lapin are favourite dishes of British travellers visiting restaurants everywhere from Calais to Cannes, and hutch-reared rabbit is for sale in all the markets throughout France.

But the meat rabbit has failed to make the short hop across the Channel and is largely unknown in Britain, where bunnies are found in pet shops and have names like Flopsy and Mopsy, or live in charming burrows on Watership Down. We really don't do rabbits for the home table, despite our love affair with French cuisine.

This is a pity, because the rabbit is one of the best ways of producing small-scale meat. The animals have a simple diet, they don't need much space and, most importantly in our crowded landscape, they don't make any noise.

'Breed Bunnies and Feed the World' was the headline in a recent New Scientist magazine, when the UN Food and Agriculture Organisation said that 'backyard rabbitries are the perfect answer to today's demand for sustainable development,' and noted that worldwide nearly a million tons of rabbit meat was already being produced each year. An information network covering 14 countries around the Mediterranean was being established to promote rabbit farming.

Commercial rabbits have often seemed to be on the verge of a boom in this country, only to fizzle out a few years later. The reasons are not obvious: there is a demand (we import thousands of tons, even though it is not a mass-market meat) and there are national suppliers of compounded rabbit feed pellets. The only inhibiting factor may be a lack of processors willing to handle meat rabbits in small numbers, but that is no reason why an enterprising small-holder could not develop some farm-gate trade or perhaps link up with other breeders to produce quantities which processors would find worthwhile.

David Blyth, of Woldsway Foods, Britain's only EU-approved processor, says: 'Our only problem is a shortage of rabbits. There is a big market waiting for them.'

It is certainly easier to produce home-grown meat for your own dinner table from a small unit of rabbits than any other form of livestock. A doe (£12) mated to a buck (£12) will cost about £1.50 a week to feed – rather less if you have your own hay and some vegetables – and will produce 30-50 meat rabbits a year, each weighing 6lb liveweight at 10 weeks – nearly 150lb of low-fat, high-protein white meat in the oven and the freezer. Even if your garden rabbitry extended to 20 females, you will still need only one male. Most of the costs in rabbit-keeping are feed and labour. If you discount your own labour and win much of the feed from the smallholding, the meat is remarkably cheap.

My father reared New Zealand White and Californian

rabbits for years after his retirement. He avoided all the slaughtering and processing stages by bartering with our local butcher, who took live rabbits and let us have packed meat in exchange. The rate of exchange varied, depending on whether the restaurant chefs around the district were into French provincial cooking at the time, but we always had plenty of good meat in the freezer, with enough left over to give to friends or swap for other produce.

Anyone contemplating meat rabbits as a smallholding enterprise will be startled by the capital costs shown on official agricultural websites. A 100-doe unit is shown at about £9000, plus buildings, but of course this is like the start-up figures shown in newspaper articles under headings such as: Why not take up shooting? Gun, £3000; Share in syndicate, £5000; Labrador, £2000; Range-Rover... and so on.

Not many of us start at the top and it is easier to breed your own stock and build your own hutches. Very little of that million tons of rabbit meat that is eaten around the world is produced from modern, artificially-lit, wire-floored cages and fed on pelleted feed. Most of it comes from peasants' stock kept in DIY hutches and fed from the garden and the road verges. Breeding rabbits is not rocket science. Start small and learn as you go along.

Looking back on the rations recommended for British rabbits in wartime, it is easy to see why they were considered an economical way to eke out the few penceworth of meat that formed the weekly ration in the 1940s:

Monday: An armful of weeds such as dandelion, ground-sel, sow thistle, vetch, shepherd's purse.

Tuesday: Fresh lawn mowings (these can also be dried).

Wednesday: Vegetable tops and trimmings, leftover salads

(but go easy on the lettuce which, despite the mythology, is not good for rabbits in large quantities, especially females feeding young.)

Thursday:	Weeds and household scraps such as dry bread.
Friday:	Hedge and tree prunings, grass.
Saturday:	Clover, lucerne.
Sunday:	Cabbage, kale.

Hay or barley straw is needed every day, with plenty of clean water.

This is the sort of diet that the Continental backyarder is feeding to his rabbits, throwing in almost everything from the vegetable patch and flower garden with the exception of rhubarb, potatoes and bulbs. It is labour intensive but inexpensive. Feeding greens and roots works well on its own, and a pelleted diet alone is also satisfactory, but alternating between greens and compounded rations does not agree too well with the rabbit's digestive system.

Rabbits can be kept in movable pens on grass. At first sight, this is an attractive option, but it turns out to be inefficient. Working with mating does, young litters and growers out in the rain is not much fun – there is much more hands-on activity with rabbits than there is with poultry.

Although rabbits are active by night, they need at least 16–17 hours of daylight to bring them into breeding condition. This is the main reason that productivity of back-garden hutches falls off in midwinter – it has nothing to do with the cold weather. A timer and a couple of 60-watt bulbs in the shelter over the hutches will work wonders. Give the does plenty of hay in which to build warm, fur-lined nests and they will breed all year round.

The nest has to be warm because the female does not spend all her time nursing the young rabbits. At first she will suckle them only once in 24 hours, a fact which makes it possible to transfer very young rabbits between breeding units at day old.

This method was pioneered about 20 years ago by French breeders, who found that day-olds showed much lower losses than animals moved at the traditional age of 8–11 weeks and they certainly had a better survival rate than rabbits shipped between breeders at four months or older. The young are put into the nest of an experienced doe in their new home and she accepts them readily, her own young having been fostered as 'extras' in the litters of other females in the rabbitry.

The tiny young rabbits – naked, blind and helpless – are remarkably resilient if they are boxed together in a group and not allowed to chill. Many litters have been transferred from the West Coast of the United States to France without any losses on their 36-hour journey.

The reason that the young need such infrequent feeding in the early stages is that rabbits' milk is more nutritious than that of any other farm animal – about 1000 calories per lb. compared with 350 calories in cow's milk.

At 10 weeks, young rabbits of a commercial strain weigh 5.75–6.5lb and are sold to the processor. Current price is in the region of 55p per pound live weight, with prices slightly higher in winter.

Before leaving the subject of backyard rabbits, it's worth explaining why they are such a strong tradition in France.

It all started with Napoleon III, who was exiled as a young man to a castle in central Europe. During his house arrest, he amused himself by breeding livestock and especially rabbits. Eventually he managed to escape to London, where he lived until 1848, and he returned to power as Emperor four years later, but he had not forgotten the rabbits. He went on to create colonies of little farms throughout France, many of them financed from his own pocket, and every smallholder had to pledge that he would keep rabbits.

In Alsace, Mulhouse and South-west France, thousands of holdings produced millions of the animals. One fur-processing plant at Le Mans was handling 18 million pelts a year and farmed rabbit became a mainstay of the rural menu. This is a deliciously French recipe:

2 rashers of bacon, chopped	quarter pint stock
2 rabbit joints	salt & pepper
2 carrots, diced	a bouquet garni
1 onion, chopped	5oz yoghurt
1 stick celery, sliced	

Fry the bacon, add rabbit joints and brown. Add carrot, onion and celery, pour the stock over, season and add bouquet garni. Cover and simmer over low heat until tender, about 1 hour. Stir yoghurt into mixture and cook without boiling for further 1–2 minutes. (Serves one).

Rhubarb: The perfect pie plant makes a comeback

The attraction of rhubarb for the hobby farmer can be summed up in one word – it's easy. The 'pie plant' is very willing to grow and it thrives for years with low maintenance. It is largely free of pests and diseases, and there is a demand for the end product.

The downside is that the rhubarb harvest demands quite a lot of labour in the fields and the plant has a brief market season in spring, even though it goes on growing all summer. And you do have to keep on top of the weeds.

The biggest point in rhubarb's favour is its return to culinary fashion after many years as a neglected fruit/vegetable. Even multi-star chefs like Gordon Ramsay, after noting that it was 'the sort of thing that dads grew in their allotments and mums ruined in a crumble', are now enthusiastic about its well-deserved comeback in the kitchen.

In Gordon Ramsay's opinion, humble rhubarb is 'brilliantly versatile' and makes a sensational accompaniment to things like scallops and lobster and has the acidity to cut through the oiliness of fish such as mackerel.

For hundreds of years, it was a plant of magic and fascination. The medicines extracted from rhubarb root were so highly prized that on the London market it sold for ten times the price of cinnamon and nearly three times the price of opium. Expeditions were mounted with the sole purpose of finding the source of this fabulous plant and learning more about its mysterious healing powers. But although pioneering botanists and physicians cracked the secrets of most Eastern spices, even today the medicinal rhubarb remains something of a mystery.

While the physicians' rhubarb of China still intrigues the experts, its Siberian relative found a place in the kitchen gardens of Europe by the 1600s and into the tart and pudding recipes in which it is still found today. Its name derives from Latin and means 'the barbarian plant from beyond the Volga'. As its origins would suggest, rhubarb is a very hardy, cold-resistant species which needs a few sharp frosts before it will set buds for its rush of spring growth, the pink leaf stalks which are the earliest of garden 'fruits'.

Rhubarb appreciates our northern sunshine and should be established on an open site. It will survive for years in partial shade, but under those conditions it never has the lush growth of a plant thriving in the sun. It is one of the most permanent of garden plants, coming up for 15 years or more from the same crown, and since it is rather intolerant of deep digging around the roots, it should always be sited in a place where it can be left undisturbed apart from some mulching and shallow weeding, and perhaps splitting off a few new crowns after 4–5 years, slicing them out with a sharp spade. Each crown taken in spring should have three or four buds and will root very quickly in its new site.

Rhubarb plants are available all year round at many garden centres, although the best time to plant rhubarb is late autumn to early winter – December is about right. The crown divisions should be planted 3 to 4 feet apart and rhubarb is not fussy about the type of soil. A mature plant of a strong variety such as Victoria can be 4 feet (1.25 metre) in diameter and 3 feet (1 metre) tall. Plant the roots with the crown bud close to the surface of the soil. Rhubarb can also be grown, slowly, from seed. However, there is little guarantee that the seedlings will come true to their parent variety.

Commercial rhubarb requires about 1,500lb of 10:10:10 fertilizer per acre. In the garden or smallholding plot, a cup of 10:10:10 for each plant in spring is recommended, although rhubarb is a greedy feeder, as its growth of lush leaves would suggest, and large helpings of manure, compost and mulch are appreciated. Do not be shy about shovelling on the muck, the response will always be positive.

Flower stalks on rhubarb look quite dramatic – towering fists of greenish-white flowers above the leaves – but their production does drain the system and they should be trimmed off as soon as they appear above crown level. Despite old wives' tales, the appearance of the flowers does not mean that the stalks have turned poisonous. The edible parts are not affected, although the leaves are poisonous at all seasons to both humans and stock. They make good compost.

During the first year after planting, the stalks should not be picked. Give the crown time to develop. One light picking may be taken during the year following planting if the plants are vigorous, and from the second year onwards all the spring stalks can be harvested. Rhubarb will tolerate a fair amount of neglect and still thrive – these are very tough plants. Established rhubarb crowns can be coaxed into early outdoor production by covering them with terra-cotta forcing pots or upturned buckets. Almost as good is a sheet of clear plastic placed over the root in late winter, just before growth starts. As pink buds appear on the top of the rootstock, cut ventilation slits in the plastic and as the leaves develop, slash the plastic to let them grow more freely.

If the harvested stalks are washed in water, they may split and will probably be unmarketable, so if they are particularly mucky, try wiping them with a damp cloth. Pulling the stems from the crown rather than cutting them, and also leaving a little piece of cut leaf at the top of the stem, will help to prevent splitting.

Clean rhubarb should be stored in a cool dry place, preferably a refrigerator. If the stalks are fully mature when they are picked, they will keep very well for a couple of days. Rhubarb can also be stored in the freezer and used all year round. Chop the stalks into one- or two-inch pieces and put them into airtight freezer boxes.

When the stalks start becoming very thin, it is time to stop picking. As summer approaches, the sticks become coarser and more fibrous, with a flavour which tends towards the sour rather than simply tart. There are reported to be varieties in Scandinavia which are harvestable through June and July; an enterprising smallholder might find it worthwhile to track these down and try testing the market for summer rhubarb to mix and match with soft fruits in their high season.

The culture of rhubarb for the sake of its medicinal root rather than the stalks was first attempted in England at Banbury, Oxfordshire, in the late 1700s by an apothecary named Hayward, who raised plants from Russian seed. This was considered to be such a breakthrough that the Society of Arts awarded him its gold and silver medals. The Banbury rhubarb is still in existence and appears to be a hybrid of at least two Chinese species. When Hayward died, he left his rhubarb plantations to the family who still grow it in Oxfordshire.

Chinese rhubarb reached Europe chiefly by way of

Moscow; and from 1704 the rhubarb trade was a monopoly of the Russian government. The roots proved difficult to preserve on their long route across Asia and this was one of the costliest of rare medicines. It was often incorporated in a purgative called Gregory's powder, composed of 2 parts of rhubarb, 6 of magnesia and 1 of ginger. The drug is peculiar in that the purgation is succeeded by constipation, so it is a one-off cure. Rhubarb root was also used to help in the weaning of infants, since its taste comes through in mother's milk, giving it a bitter taste.

The botanical source of Chinese rhubarb is a mystery even today. The roots produced by wild species under cultivation in Europe do not have the veined markings present in the best specimens of the Chinese root. Their cultivation apparently remains shrouded in secrecy.

The modern home of British commercial rhubarb for the kitchen is just south of Leeds, where the M62 crosses the M1. This is rhubarb's Golden Triangle. If you look at the map, this region between Rothwell, Morley and Wakefield, centered on the village of Carlton, looks as if it should be the setting for smokestack industry, but its warm, candlelit sheds are the heartland of traditional rhubarb forcing. It produces those rosy pink stems with pale yellow leaves whose arrival in the greengrocer's baskets in February is one of the first heralds of spring.

The combination of delicate texture and brilliant tart flavour is a best seller. After flirting with all sorts of long-haul foreign fruits, British cooks are returning to the long-neglected rhubarb of the allotments. Production in the Golden Triangle has doubled in recent seasons. Rhubarb is seen as tasty and also healthy, with a reported ability to lower cholesterol levels – the mystery plant of the Orient is having a renaissance.

In the kitchen, it adds a zing to flans and tarts like nothing else from the garden. It lifts plain strawberries, raspberries and apples, making them even tastier. Rhubarb also makes a zippy sauce for chicken, venison, halibut and salmon. Chefs now load it with singing superlatives.

Sheep Flock: Good looking – and adding value

Sheep do look good in the park. An estate agent once told me that a pretty flock of sheep in the foreground of a picture of a house could boost the price by £5,000.

'If I was selling a country house – ten acres, paddock and orchard, stables, all that – I'd probably hire some sheep,' he said. 'The spotty ones would be best. Might even add 10K.'

The 'spotty ones' he had in mind were Jacob sheep, just perfect for the park. And it is even possible to pretend that these attractive creatures, which so enhance the view, are being kept in an honest endeavour to make a profit.

For the owner of a country house, the paddock farming enterprise can have a Cinderella status, neither profit-making nor an out-and-out luxury. With a few sheep around, it can at least have an income and hold out the possibility that one day a prize-winning shearling ram from the orchard is going to make £1000 at auction.

Short of getting a Wild Animal licence, you are never going to be able to keep anything more dramatic and handsome than a ram of one of our native breeds.

Sheep and Man have been living together for about 8000 years now, and we get along very well. Sheep adore their shepherd and make easy pets, needing the simplest of care: a stockproof fence, plenty of good grass, water, a windbreak and a salt lick. On small acreages, they will need extra hay in winter and perhaps a handful of concentrates. They are not difficult to look after and have the great virtue of being not only hardy but also tame and easy to handle.

To your accountant, you can affirm that you are keeping sheep for meat, wool and efficient management of your grassland. After all, 45 million sheep are kept in this country in expectation of making a profit. You do not need to explain that the true rewards are the wonderful moments when new-born lambs take their first hoppity-skippety outings with their mothers in the meadow, or when a proud ram poses for a photograph, turning his head into profile, making sure you get the best view of his horns.

My old *Agricultural Note-Book*, published nearly a century ago, says that 'sheep of all kinds are normally fed on grass only, summer and winter'. If you are keeping them as ornamental livestock, with no rush to fatten or produce early lambs, that is still true. And the more ancient the breed, the truer it is. Soays, Manx and my own rare Castlemilk Moorits – all northern short-tailed primitives – are more likely to suffer if overfed than they are if left to their own devices on natural grass. And they do eat weeds, including ragwort.

Most of Britain's numerous breeds were developed in harmony with their local conditions. Wherever you are, from Shetland to Portland, there is a breed which does well where you live – and it may well be grateful for your interest, because several breeds are quite low in numbers and come under the cloak of the Rare Breeds Survival Trust (RBST) (024 7669 6551). The National Sheep Association (NSA) (01684 892 661), lists 84 breeds, with addresses for breed secretaries, and explains the legal requirements for registering a flock. The NSA website is a good starting point.

Not all of those 84 breeds are original British types, and you might find such immigrants as the hornless, lop-eared Est a Laine Merino more attractive. If your dream is Mediterranean rather than Hebridean, you could have a tight little flock of southern sheep sheltering under the evergreen oaks and gum trees in your sunny park. They look wonderful and it's easier than buying a villa in Majorca.

There is something for every paddock owner in that long list of seven score breeds. There are big kinds which are easily kept within bounds and little ones which like to burrow and jump. There are meaty ones and woolly ones, black, brown and white ones, and of course the spotted glamour girls. There are rough-woolled Herdwicks from the mountains and smooth Hampshire Downs which look more like furry toys than farm stock.

The splendid Wiltshire Horn, one of the best-looking of all our natives, has a coat of hair rather than wool, which does away with the need for the annual circus of shearing the fleece – the one time of the year when the small-scale sheep keeper is likely to need professional help.

There are fashionable breeds for the paddock, led by the Jacob and the Hebridean, both of which have been parkland decorations for centuries, especially in their four-horned variants. And currently the colour forms of Welsh sheep, such as the black Balwen with its white blaze, socks and tail-tip, are joining the long-tailed Black Welsh Mountain as a smart breed. All are thrifty and the lambs 'hit the ground running'.

The traditional time to buy stock is at the great sheep sales in September. Rams go on with the ewes in November to produce lambs 21 weeks later, at the end of April. First-timers might like to check out the breeds by visiting collections such as the Cotswold Farm Park, near Stow-in-the-Wold, Gloucestershire (01451 850 307), and then contacting local breeders direct via the RBST or the NSA.

The end product: a pullover from your own wool, maybe some tweed, and the best roast lamb.

Soft Fruit: The Next Big Thing in the garden

July is the high season for soft fruit. If you already have productive raspberries, blackberries and loganberries on your home acres, these are the weeks when the crop goes into the jam pan or the freezer. And if you are thinking about establishing a few canes in your kitchen garden, it's the time to shop around, do some taste-testing and decide which varieties you really like.

The attraction of cane fruit is summed up in the nurseryman's catalogue: 'Vigorous, trouble-free, up to 20lb. of berries per plant when mature.'

Some of the latest easy-care varieties, unfortunately, have rather bland berries, lacking the zing of traditional raspberries and blackberries, so you have to do the sampling. Pop into a few 'pick your own' farms and compare the qualities of the fresh berries, which are very different from the fruit in a supermarket punnet. None of these berries travel well – they are extremely perishable even under cool conditions, and they bruise easily – so you have to track them down on the farm.

Hybrid fruits are definitely the Next Big Thing in our own

hobby-farm garden, so we have been doing a lot of tasting and making some tricky decisions. Our new blackberries, tayberries, loganberries and autumn-fruiting raspberries are all in their first season of growth, zooming up their posts and looking healthy.

Most of our hopes are pinned on the tayberry, described by everyone who knows about fruit as a 'sensational hybrid' between blackberry and raspberry, and also on the loganberry, a longer-established variation on the same parentage. We particularly like the comment by the French authority who wrote: 'These two species are very interesting because they are never sick."

The tayberry was introduced by the Scottish Horticultural Research Institute just over 20 years ago and has been a success on both sides of the Atlantic, making considerable inroads in the traditional berry-growing regions of eastern Scotland and in Oregon. One American review is headlined: 'The best hybrid berry yet?' There is a question mark after that line, but the verdict seems to be: yes, the best yet.

The hardy tayberry is a cross between the Pacific Coast blackberry aurora and a modern raspberry variety. It develops long biennial canes, growing to 6ft or 7ft in one season, usually tied to wires between posts. These thorny canes flower and fruit in the following season, after which they are cut down and replaced by new growth from ground level. Many commercial growers tie the new canes in bundles as they develop. After the harvest is over, the old canes are cut and the new ones spread out on the wires, a cycle which can continue for 15-20 years on a well-manured site.

Tayberries are 1.5–2in. long, reddish to purple in colour, sweet and aromatic, with much of the tangy aftertaste of a raspberry – slightly less acid than a loganberry. Fruit is picked in July and August and is particularly suitable for freezing, although it is also recommended for jam, canning and bottling.

Like the loganberry and blackberry, and unlike its raspberry parent, the tayberry retains the white plug or core of the berry when it is picked. If you are organically-inclined (or lazy) your fruit may be invaded by the raspberry beetle's tiny maggot, which lives in the plug.

As one of my old cookery books notes: 'The chief precaution to be taken when making loganberry jam is to get rid of the grubs...letting the fruit stand for about quarter of an hour in a weak salt solution, when the grubs will come out and rise to the top of the water, when they can be removed. The fruit must then be put in a colander and cold water poured over it to remove all trace of the salt."

The loganberry has a much longer history than the tayberry and, despite those grubs, has been popular ever since its introduction by Judge J.H. Logan of Santa Cruz, California, in 1880. The parents were probably a Texas Early blackberry and red Antwerp raspberry. The fruits are smaller than those of the tayberry, but it continues to be popular because it is so willing to grow, carries a heavy crop and makes great jam.

Your local nurseryman might even stock some of the exotic-sounding cane fruits which have been strongly promoted over the years: boysenberry, phenomenal berry, newberry, laxtonberry, king's acre berry, lowberry, youngberry or veitchberry. Most of these make colourful and tasty jams, but only the tayberry – and in very cold places, the tummelberry – looks like finding a permanent place in our gardens and smallholdings.

Virtually all of these are variations on the raspberry-blackberry hybrid, using European and North American species. The blackberry of America's Pacific Coast has also been crossed with our native bramble to produce the modern garden blackberry, one of the few fruiting plants that can be expected to thrive on a north-facing wall. The upright, short-caned varieties of the garden blackberry, such as Loch Ness, are described as 'just about the lowest maintenance backyard fruit available.' A strong recommendation, because the berries are full of flavour and the canes yield up to 10lb of fruit a year.

Another advance in soft fruit is the autumn raspberry, which produces berries on the current season's cane growth rather than wood developed the previous summer. These primocane varieties, such as Autumn Bliss and Joan J (both from S.E.Marshall and Co., Freepost PE787, Wisbech, Cambs. PE13.2WE, 01945.466.711) produce their berries a bit later than the traditional high-season raspberry, from early August onwards. They have the great advantage that the entire growth can be cut and swept away after the last berry has been picked some time in October (or even November in a mild season), which really does help to keep pests and diseases under control. Their only disadvantage is in the flavour department: they are rather bland.

So what are we going to do with our enormous quantity of soft fruit when it all ripens next summer? The answer seems to lie in the freezer – most of the berries freeze very well – and also with the old Kilner jars and Le Parfait bottles in the loft. Those glass jars are about to be fitted with new rubber gaskets and enjoy a fresh lease of life.

An easy way to preserve the larger and firmer fruits such as Victoria plums is to put the fruit into the jars in the oven at 100°C for about an hour, standing the glass bases on a wooden board to prevent cracking, and cook until the plum skins are just splitting. Top up each jar with boiling syrup (8oz of sugar to 1 pint of water) and seal at once.

The loganberry jam recipe is just as simple: 6lb of loganberries, 6lb of sugar, no water. Boil berries in the preserving
pan until reduced by about one-third, add sugar and boil
briskly. As soon as the jam sets when a sample is tested on
a cold saucer from the fridge, pot and seal. Over-ripe
loganberries may need a tablespoonful of lemon juice to
help the set.

Footnote: A thriving thicket of thorny blackberries or
tayberries along a kitchen garden fence is a useful deterrent
against intruders.

The Fig: Sun-loving fruit is moving up the map

 The fig is a sensual Mediterranean fruit with a history as long as civilisation – and perhaps a promising future in our warming British Isles. The day may be coming when you will be able to take a walk around your English smallholding and pick ripe figs like you have done so often at your holiday villa in Greece or Spain.

Beautiful, decorative, producing 'the world's most luscious fruit', the fig in its native part of the world is the perfect garden tree, bearing two or even three crops of fruit each season. It is surprisingly hardy – in many places on the Continent it withstands minus 20°C – but it would prefer our summers to be just a bit longer and a bit hotter to ripen its shoots and its fruit.

For most of us in this country, a productive fig has been a tub-grown plant which could be wintered in a cold shed for a few months then given a warm-up in early spring to get its fruiting season off to a cheerful start. And figs do thrive like grape-vines when grown in tubs or large pots in a sheltered courtyard. A 3ft fig is happy in a 14 inch pot and will do particularly well under unheated glass.

As a permanent fixture in the orchard, however, the fig has always promised more than it produced – more figlets than figs – unless you are within reach of London's expensive microclimate, boosted by thousands of central-heating systems. The Brunswick figs on the Trafalgar Square frontage of the National Gallery show how well this exotic-leaved tree grows under London conditions, even when drastically pruned.

The good news is that continental plant breeders have been working on varieties which will extend the fig's range. 'Thrives south of London' may soon become 'thrives south of York'. Like the grape, the sun-loving fig is moving up the map, especially if it can be given the warm backing of a wall.

There is a lot of genetic material for the plant breeders to work with. More than 700 distinct varieties, with a colourful array of fruits in yellow, purple, green, red, brown and black, grow around the Mediterranean. Turkey and Greece are the major world producers, with California in third place, followed by Spain and Portugal.

The Roman writer Pliny said: 'Figs are restorative. They increase the strength of young people, preserve the elderly in better health and make them look younger with fewer wrinkles.' And so the Romans took their fig trees with them throughout the empire so that they could keep their wrinkles at bay.

Figs were probably one of the first fruits to be dried and stored by man. Under arid conditions, they dry themselves on the tree. These dried fruits, which are a wonderfully transportable luxury, have travelled everywhere that Western man has penetrated. Archaeologists digging in British cesspits from Roman times onwards invariably discover fig seeds, although it is difficult to distinguish whether these are the seeds of locally grown fruit or whether they had crossed Europe by pack-horse to grace the table of some baron or bishop.

In commercial fig groves, the fruits ripen to maturity and fall to the ground to complete the drying process. Harvesting is simple, you just pick them up. The fruit can be then be stored in its dried state for months without deteriorating. Under British summer conditions where 'dry' is a rare condition, the fruit can be picked fresh when it colours and wilts on the tree – a fig is ripe when it sags on the branch and begins to split at the narrow end – and if you are faced with too many to eat at one time, the surplus can be put in the freezer.

The Holy Roman Emperor Charlemagne attempted to introduce the fig to the Netherlands more than a thousand years ago, it is said, but he was unsuccessful because the tree could not ripen fruit during the short northern summer. It is possible that Charlemagne's figs depended, like their wild ancestors, on pollination by specialised wasp species found only in warmer southern climates. These enter the embryo figlet and struggle about inside, coating themselves in pollen before moving on to another fruitlet to fertilise the seeds and lay their eggs. Modern figs are all self-fertile female trees and need no wasps. This has two benefits: (1.) a single tree can produce fruit, and (2.) you don't find yourself munching on a maggot.

Although Charlemagne's efforts failed, today the Belgian nursery of Penninckx offers nearly 30 varieties and promises ripe fruit from all of them under garden conditions 'in our not-so-warm summers' – and even two crops from some cultivars. Conditions in Belgium are very similar to the South-East of England, where a picking of figs from the popular Brown Turkey variety has been possible every summer in recent years, and we can look forward to more of these Low Countries varieties reaching our nurserymen.

The fig is a curious tree because it is most fruitful when it is struggling to make a living. Traditionally, wall figs are planted with their roots in restricting boxes of brick or stone to prevent leaf growth and Californian experiments have recently shown that the crop can be doubled by narrowly ringbarking the trees in commercial orchards.

If your fig is over-enthusiastic, producing lots of big leaves but not many fruits, one answer is to dig deeply in a semi-circle around the tree (5ft from the trunk and about 18 inches deep), cutting the roots. Repeat this every two or three years, restricting root growth in a full circle, until it gets the idea and turns into a fruit tree instead of a rampant ornamental shrub. Another good reason for such harsh treatment is that luxurious, soft growth is more readily damaged by frost.

A fig tree matures in three or four years, although it may start offering some fruits from the first year after planting out. Few of these figlets will develop into anything worthwhile in the first couple of years. A well-established fig tree will grow more than 30ft high and wide against a south-facing wall. Unlike most exotic fruits it is surprisingly free of diseases and browsing pests do not like the bitter white sap.

Despite all the nice things the Romans said about the fig, in today's Britain it bears a health warning: the milky sap may cause skin burns in combination with sunlight.

If you experiment with a few fig trees in tubs, they can be overwintered in their pots in a cold but frost-free shed or cellar. Make sure they are fully dormant before putting them away for the winter by leaving them out for the first touches of frost, which will make sure they have closed down until spring. Knock off any half-grown figlets on the branches because these are very unlikely to develop properly. The following summer's fruits are present as very tiny buds in the leaf axils.

Overwintering figs should be kept cold, dark and almost dry. When the sun begins to have some warmth, about the end of March, bring your figs out to a hot and sheltered spot and give them a lengthy soaking, followed a few days later by a feed with liquid fertiliser. They will need regular watering or the embryo figs will drop off almost as soon as they start to swell.

The problem with the British climate, of course, is that spring makes promises then changes its mind. Sunny days are followed by freezing winds. If that happens, be prepared to throw a blanket over your fig trees for a night or two until the warmer weather comes back. The main setback to figs comes during April and May. A couple of frosts as they are breaking into leaf can shorten their season dramatically.

Although there are hundreds of varieties around Europe, and as many more in California, the list offered by nurserymen in England is short: Brown Turkey, Brunswick, Black Ischia, White Marseilles and Bavarian Violetta would complete the catalogue for most of them.

Brown Turkey, which may actually have originated in Provence, is self-fertile and produces large fruit with a bronzed purple-brown skin and rich amber-pink flesh. This fig is known for reliably good flavour when eaten fresh and for its hardiness, including a strong ability to tolerate salt water. It starts bearing fruit at an early age, sometimes in the first year after being planted out.

Figs are straightforward to propagate. Four-inch long ripe cuttings with a heel will strike readily in pots of soil which have some added peat (10 per cent) and sand (20 per cent). In a warmed propagator, these will root in 2–3 weeks.

Throckmorton's Coat: The challenge to make your own woollens

At five o'clock on the morning of 25 June, 1811, Sir John Throckmorton, Bt, arrived in front of Greenham Mills in the Berkshire village of Newbury, accompanied by his shepherd and two sheep. The timekeeper checked his watch and the race was on, with 1000 guineas at stake.

Sir John, the squire of Buckland, near Newbury, had wagered that he could take the fleece off a sheep's back at dawn and sit down to dinner that night in a fine cloth coat made from the wool.

Even in an age of extravagant wagers, it was a risky bet. The processes of shearing, washing, carding and spinning on treadle-driven wheels, followed by weaving on a hand loom, scouring, fulling, tenting, raising, dyeing and dressing all took time and skill and it was eleven hours before the cloth could be handed to a tailor for cutting, stitching and pressing into a hunting coat.

At 20 minutes past six o'clock, mill-owner John Coxeter, who had organised the spinning and weaving skills to help Sir John win his bet, handed over the damson-coloured, long-tailed coat, 'well woven and properly made' to the squire in front of a cheering crowd of 5000 people.

And at eight o'clock, Sir John sat down to dinner with 40 gentlemen, richer by 1000 guineas.

The two sheep were roasted for the spinners and weavers. The mill owner donated 120 gallons of beer and the celebrations went on all night. The Agricultural Society struck a silver medal to mark the record and the coat still survives, to prove that the workmanship was meant to last.

The Throckmorton Coat is on display at the family's seat at Coughton Court, Alcester, Warwickshire, to which they moved after leaving Buckland, but it no longer holds the record, which is claimed by a similar coat now on display at the West Berkshire Museum in Newbury and which was made in 1991, in one hour less than the 1811 time.

The record for a knitted sweater rather than a woven cloth coat, is very much shorter – just 5 hours 9 minutes, achieved by a team from Shetland in 1997. They were competing against teams from several other countries, including Australia, New Zealand, Japan, USA, Canada and all parts of the British Isles, in the 'Back to Back Wool Challenge'.

The rules for taking the wool from the sheep's back to the man's back in this challenge are straightforward: just eight members in the team – one blade shearer and seven hand-spinners and hand knitters – with up to seven non-electric spinning wheels and one sheep, any colour, any breed. The sweater or jumper is knitted to a standard pattern.

That's the fast way of doing it, but how does the hobby farmer with a small flock of sheep go about producing a pullover or some tweed from his own wool at reasonable cost?

Richard Martin, of Cotswold Woollen Weavers, at Filkins, near Lechlade, Gloucestershire, told me: 'If you have only a couple of fleeces, you have to to deal with them by hand – a spinning wheel and carding bats. For larger quantities, you move up to industrial machinery, however small.

'So if you have only a few sheep, your best move is to go and enrol in evening classes and learn how to spin. Or get in touch with your local spinners and weavers guild – there will undoubtedly be one – and ask if anybody would like to spin the fleeces for you.'

He added: 'The alternative is to use your two fleeces to mulch your garden.'

But assuming that you are keen to wear your own wool, he says his own mill would need to process 150 kilos, about 50–60 fleeces, as a minimum to make it worthwhile, although he does a lot for co-operatives who make up the weight by pooling their clip.

And would the wool of your rare-breed sheep be acceptable? Richard Martin says that technically only one thing makes a difference and that is if the wool comes from a very long fleeced breed – Leicester, Cotswold, Lincoln, Wensleydale – which would require worsted spinning rather than wool spinning.

He explains: 'In woollen spinning, the wool is carded and the tangles are taken out, but the fibres themselves lie higgledy-piggledy relative to each other. In worsted spinning, the wool is combed so that the short fibres are taken out and the long fibres that are left lie parallel to each other, creating a harder, denser yarn.

'That combing process requires quite a large feedstock to make it work, so you might be talking about half a ton or even a ton to do it commercially. So those long fibres are difficult to work, but with anything else, from medium lengths downwards, it does not make much difference.'

On the subject of undyed natural wool, he says: 'Traditionally, we have done a lot with coloured breeds – Jacobs, Herdwick, your Castlemilk Moorits, and especially Ronaldsays for the owner of North Ronaldsay. In the case of Jacobs, it is possible to separate the wool out into dark and light shades, weaving checks without any dyeing at all.

'People who have white fleeces may not want a completely white product, in which case we might make a travel rug, for example, in a Tattersall design with a white ground and a yellow or green check, providing the one or two per cent of coloured wool needed.'

Richard Martin says that there is increasing demand at the 50-fleece level as farmers show more interest in adding value and selling further along the chain, although he warns that it is relatively easy to make products but relatively difficult to sell them.

'There is always a honeymoon period,' he says. 'They ask how much it costs and I tell them that it will be £4 or £5 a kilo for spinning the yarn and £10 to make a travel rug. They will have a very nice travel rug for £15. That's great, they say, I can sell that for £40.

'They forget that you have to put up a couple of grand to buy the rugs, a big capital investment. And you have to have a reasonable idea about where you are going to sell them. But people do, of course, and lots of our customers come back, year after year.

"There's a "vanity publishing" side of this spinning and weaving business. Flock owners have enough wool spun to make pullover because they can say: This is my wool, from my sheep.

'Then there is something a bit more ambitious, which is where we come in.

'At the vanity end, the principle is the process: the fun is in doing it. The fact that you get a pullover is almost irrelevant. Our point at Cotswold Woollen Weavers, on the other hand, is the product. You can put a price on it and you can sell it.'

The ultimate vanity-publishing project is an estate tweed worn by lairds, stalkers, keepers and ghillies on Scottish estates. Like tartans, they are the subject of much folklore and history, but unlike tartans they are not usually linked to a family but to a place, whether it is Aberchalder or Altnaharra, Wemyss or Wyvis.

The specialists are Johnstons of Elgin, whose designer John Gillespie told me: 'There is still a niche market for estate patterns. We can design a tweed for anyone who is aiming to buy 60–120 metres.' The mill will work with ideas and colours suggested by the estate owner or can amend an existing pattern.

These estate tweeds have been the origin of many of today's fashion standards, such as the 'gun club' pattern which was created in the 1840s for Coigach, near Ullapool in the North-west Highlands, and the stylish Glenurquhart, first sketched in the mud outside a hand-weaver's door by Elizabeth MacDougall of Lewiston. Today they are more likely to be made in fashion fabric or cashmere.

And the ultimate is a beautiful estate tweed from your own rare breed in the park. The Earl of Mansfield's flock of Jacob sheep in the grounds of Scone Palace at Perth provides the wool for his unique tweed, woven by Johnstons.

Tree Planting: The golden rule – get on with it

There are two stages in planting amenity trees around the hobby farm. The first involves planning, landscaping, buying saplings and putting them in the ground.

The second involves doing it all again because rabbits and deer destroyed the first lot.

Your personal list of tree-killers may include voles and vandals, perhaps even thoroughbreds and half-breds, but wherever the hobby farmer has planted trees, there has always been something out there waiting to nip the idea in the bud.

Which is the reason why we should all say a quiet 'thank you' to the man who invented the tree shelter. Nearly 10 million trees have had a better start in life since forester Graham Tuley first experimented with the idea in his Hampshire garden in 1978. His now-ubiquitous plastic tube not only gives young broadleaf trees protection against browsers but also boosts early growth, sometimes more than doubling the speed with which the trees put on height and girth.

A tree shelter costs more than the tree itself and it is usually put around broadleaf trees because conifers show much less growth benefit, although they do need browsing protection. Tubes are almost always a good investment, and the way in which they signpost the infant trees among tall grass and weeds speeds up the application of the annual dose of herbicide – another lifesaver, giving them room to root and breathe.

At the planning stage when thinking about new trees, the main point to remember is that a tree which has taken 10 years to grow can be cut down and chopped up in 10 minutes, so it's easy to change your mind. Get on with it.

The usual regret with tree-planting is that you didn't do it 20 years ago. And decide at the beginning whether you want to fence off all your trees en bloc or to protect them individually.

In cases where the smallholder is planting coppices or shelterbelts, he may want to deer-fence the entire block, then put small tubes around individual saplings to encourage fast growth. Or if the plan is to establish some specimen trees in the field in front of the house, hefty tree guards might be needed to hold back horses and cattle from each sapling in its nursery tube.

When Humphrey Repton was sketching his landscape ideas for England's great estates in the 1700s, he always showed how fencing would eliminate the browsing line – where cattle in the park had destroyed all the lower branches of the ornamental trees – and could create much more attractive spinneys. Some people like the browse line and the cattle in the shade, but if you don't, be prepared to build permanent fences because once the lower limbs are gone, the trees will never sweep back down to ground level again.

If you were lucky enough to have an appropriate education, you know already that there is a formula for deciding whether it is cheaper to put up a perimeter fence or to buy individual tree shelters and protection. It looks like this: (FxP)/(NxC). F is the per-meter cost of fencing; P is the perimeter (total length of fence) in metres; N is the number of trees to be protected per hectare; C is the cost of each tree tube/guard. If the answer is less than 1, a fence is more cost-effective.

Whatever the sums may suggest – and the formula probably indicates that any new woodland plot over a hectare should be fenced – there are lots of variations on the tree shelter or tree guard which can pay their way. Some give browsing protection only, some help the little transplant to get a good start in life.

The simplest is the plastic spiral, wound around the stem of the young tree and its supporting cane. This gives bog-standard protection against rabbits and voles, but deer soon learn that the the spirals signpost the way to a better snack. Taller guards of plastic mesh or wire mesh on sturdy posts are useful in preventing damage by deer, hares and rabbits, and even heavier mesh can be wrapped around large trees to defy the teeth of horses and cattle, but none of these encourage growth.

So the belt-and-braces approach does pay, with big stock kept at a distance by posts and Rylock, anti-rabbit netting added on the lower part of the fence, and plastic tubes cossetting the individual saplings. At the planning stage it looks like overkill, but in the end the would-be woodsman is glad he took the trouble.

Tube-like shelters are most efficient when they are used on transplants about one or two handspans tall – say 8–16in. or 20–40cm. Shelters 60cm high protect against rabbits and hares, 1.2m against sheep and roe, 1.8m (6ft) against larger deer and 1.8m heavy-duty tubes against horses and cattle.

In sheep-grazing paddocks, rams greatly enjoy the noisy thump and springy rebound when they practise their fighting technique against tree shelters. The standard advice is to provide them with other posts as an alternative, but nothing is quite as exciting for head-bangers as the plastic tube, so be prepared to string out some wire defences. Even a single wire tends to confuse our tups and they then turn their horns against the substitute posts.

All the publicity material about tree tubes says that they are biodegradable and will soon break down and rot away. In practice, we have found that they outlast the posts hammered in to support them and can be used for a second or third crop before they split.

Even longer lived are the strips of nylon tights which were used to tie up our early saplings to their posts – a 'handy hint' from a gardening magazine. Fifteen years later, mysterious collapses of well-grown maples were found to be due to hidden strangulation by the tights, deeply overgrown by the bark but still unrotted and weakening the apparently sturdy trunk. Make sure you cut the tie as the tree grows.

When we were putting in our first ornamental trees in the paddocks, it was tempting to buy the standard trees on offer in local nurserymen's yards, because it seemed obvious that a 6ft high tree is going to give a faster result in the landscape than a 15 inch transplant. However, some 20 years of experience in battling our gales and cool summers has shown that the little 'un in its warm tube is usually a better deal than the big 'un. It catches up and then outstrips the larger tree, with few exceptions.

There is a tricky stage when the tubes are removed and the young tree is exposed to the wind for the first time. Shelter-reared saplings are tall and apparently strong-stemmed, but they have not been toughened by exposure and can go down like ninepins on an exposed site.

A re-staking exercise is often needed – ideally with stakes which keep the lower part of the tree firm while allowing the top half to thrash about and learn how to live with the wind. On a small acreage, this kind of second-stage help is feasible, although real foresters just laugh at such a fusspot approach; they prefer to leave the tubes in place until they rot, gradually exposing the tree to the rigours of real life.

Damage by the wind is a constant theme here, and one that may not affect those lucky enough to have a sheltered, sunny park around the house. A notable lesson has been that tubes are more wind-resistant than the square-section plastic shelters; I grudged the hours spent in trying to invent some kind of spacer that would prevent the square ones from blowing flat and crushing the tree inside.

And silly little canes are a waste of money. You need to make a hole with a crowbar and knock in a sturdy post. If you don't, you will soon be looking out at a drunken parade-ground of leaning tubes and wishing you had done the job properly.

One obvious question: why bother with protection at all? Why not plant trees or shrubs which rabbits and deer will not eat?

Unfortunately, the list is a short one (although longer for herbaceous plants in the garden) and rabbits do have a nasty habit of nipping off plants for spite even though they have no intention of eating them. I recall a neat row of newly-planted Leyland cypress, staked but not tubed, which rabbits made a special 200-yard expedition to destroy over-night. Having massacred 50 trees, the pests never bothered to return to that paddock again.

Rabbit-resistant trees include alders, elder, eucalyptus (usually), birch, holly, laburnum, yew (poisonous to

domestic stock) and the Corsican pine. Shrubs which
usually get a grip before the rabbits can damage them are
laurel, leycesteria, lonicera, buddleia, berberis, the rugosa
roses and *Fuchsia magellanica*. Deer will damage all of
these, however – almost anything in the the woods may be
thrashed to ribbons by a roebuck in a possessive mood.

Turkeys: The heritage bird for the paddock

By early August, Britain's commercial turkey units are hatching the chicks which will grow rapidly to fill the supermarket freezer and grace the Christmas dinner table.

Old-fashioned farmyard turkeys – the pre-1950 types that American breeders call 'heritage' varieties – are taking rather longer from eggshell to oven, probably nearer 30 weeks than 20. But their pound-for-pound value has soared by comparison.

Standard bronze or black turkeys, especially if they can claim to be free range and organically fed, now command a big premium in the marketplace. In recent winters, discerning customers have been paying up to five times the price of a standard frozen bird from the supermarket for a farm-fresh Bronze.

The heritage turkey is the ideal bird for the acres of paddock and woodland around a country house – almost as spectacular as a peacock, with the added virtues of being quieter in the morning and tastier on the table. There is a steady demand for young poults in high summer and for quality table birds in December.

Hatching eggs also find a market at £1.50–£2.00 each, buyer collects. We have one breeding pair of old-style

Bronze turkeys and the hen laid steadily from the first week of March. She took a break to sit on a clutch and after her day-old chicks were taken away to be reared by a silkie hen, she laid again right through to into the autumn, a total of more than 120 eggs. The eggs keep well at room temperature and lose little in hatchability up to 14–20 days.

Traditional breeds of turkey do not need artificial insemination, unlike the highly-bred, broad-breasted whites of the commercial turkey world. In fact, they are particularly simple to deal with in this respect, since a single successful mating is all that is required for the fertilisation of an entire clutch.

Hidden away on the experimental farm of a North American university, there is even said to be a strain of the Beltsville small white variety in which the females are capable of producing fertile eggs without any males being present. These chicks are all male. If mated with toms, the hens produce the usual 50:50 male-female ratio.

As liberty birds, turkeys of the traditional breeds are truly hardy. The wild species of North America is an adaptable gamebird, found in Central America and the most arid bits of Texas, through the hot swamps of Florida and northward to the bitter winters of the country around the Great Lakes.

When you see turkeys roosting high in leafless trees on a sub-zero night in northern Britain, it is difficult to believe that those unfeathered red heads will survive unfrozen, but the birds are bone hardy and don't seem to notice the cold. They come for their corn in the morning as bright as if they had spent the night in the rafters of the barn.

Younger turkeys, on the other hand, are quite delicate until the age of six weeks, and particularly sensitive to damp grass and steady rain if they can't find shelter. A turkey rearing pen benefits greatly from some sheets of clear plastic roofing to let the sun in and keep the wet out. Once the poults have 'shot the red' – the stage at which the red skin of the neck shines through the moulting chick down – they are as weatherproof as any pheasant.

A family party of turkeys is particularly charming in summertime, working its way with conversational 'quit' and 'tic' noises through the grass at the edge of a wood. Sadly, they are no more foxproof than young pheasants or guinea fowl, but they do seem to outgrow the killing power of buzzards and sparrowhawks at a much earlier age. They are ready to fly up to roost before six weeks of age.

It is truly unfortunate that the word 'turkey' has come to be associated with failure, and that the character of the bird is typically seen as nervous, stupid and charmless. I have always found them to be full of character, usually calm and endlessly curious about people (although old males are sometimes inclined to be aggressive). The chicks are a delight to rear, especially when the month-old toms start to practise their display routine, dragging tiny wing feathers along the ground, fanning their tails and putting their chins on their chests to show off like their heavyweight fathers, all this while their heads are still in hatchling down.

Turkeys are much more interested in human activity than most poultry. I came across a painting the other day, showing pigeons and poultry being fed in the courtyard of a Scottish castle, with children throwing crumbs from a first-floor window. The artist, J.H. Lorimer, had caught the moment exactly, with the bantams concentrating on the bread on the ground while the black turkeys all gaze up inquisitively at the children leaning on the windowsill.

Free-ranging turkeys – the birds now known by the promoters of tasty meat as 'pasture fed' – are quite cheap to rear during the summer months. Once they get beyond the

starter crumb stage they find much of their own food –
insects and greens – and need minimal topping up with
grain and bread scraps to keep them tame and hefted on
their home patch. However, you will have to keep an eye on
them if you live near oak woods because acorns are the
natural food of wild turkeys and their domestic descendants
will wander far in search of this delicacy, so far that they
may forget the way home.

Stately turkeys look magnificent at liberty and you can
pick from a range of 'colourways' to suit your taste. The
most distinctive are blacks, black-laced whites, bronzes and
reds. Turkeys do not actually come in many strong colours
–once you get beyond black and white there are various
shades of grey and brown-red, some laced with black, some
unpatterned – but they do have a wonderful selection of
colourful names: Sweetgrass and Auburn, Blue and Slate,
Chocolate and Crimson Dawn, Buff and Lilac,
Narragansett and Royal Palm, Chestnut and Fawn,
Nutmeg and Oregon Grey.

Most of these were created in America, although the
blacks were imported from Mexico as a ready-made
domestic variety by the Spanish around 1500 and were
well-known in England only 20 years later. The Pilgrim
Fathers actually took Norfolk turkeys with them to New
England in 1620.

There are so many varieties that you can probably find
one to call your own. Or you might like to create one and
give it a fancy name – the Royal Sussex or the Old Vicarage
blonde. There are all sorts of possibilities for the keen
amateur breeder, among them the development of sex-
linkage, which is currently found only in the Auburn variety.

The most eye-catching birds are the strongly patterned
types, especially the breed called Royal Palm in North

America and Crollwitzer in this country, in which every white feather is edged with black.

The Bronzes have a brilliant coppery sheen on their body feathering. In all varieties, the adults have bare red and blue skin on the head and neck, particularly bright in the males, which are known as toms or stags. If you want a memorable creature at liberty about the place, it is hard to beat a Norfolk black or a black-laced White turkey. And all varieties look particularly good in the snow; they are not just fair weather friends.

The breeds show little variation in shape and all are immediately recognisable as turkeys, although the commercial Whites, out of sight in their rearing sheds, are massively broad-breasted. As one breeder of heritage turkeys told me: 'On the table, a factory-farmed White looks like a plastic rugby ball. An old-fashioned Bronze looks more like a bar of Toblerone!"

Not only showing more breast-bone, these older breeds have much more flavour, harking back to the 'turkey well drest' of the Tudor court feast. The motto of the slow-food enthusiasts is 'eat less, enjoy it more". And they suggest that you find a pre-war cookery book to learn how to slow roast your bird, keeping it moist and retaining all the flavour.

Much of the flavour comes from the fact that heritage turkeys are considerably older by the time they reach the Christmas table than the mass-production birds and they have firmer flesh. Their diet is more varied and they are much less bland on the plate, with more than a hint of game.

If you want to hark back to an earlier era, putting your money where your tastebuds are, you might like to shop around in your local farmers' market for traditional turkeys. They are certainly easy to order on the internet if you are in the market for a 10kg bird (feeds 18–20 people) at around £60-£80, including delivery.

Celebrity chefs agree they are worth the money – 'the meat cuts thin, like a wafer, without crumbling' – and they have been ordering regular supplies from upmarket poulterers such as Fortnum and Mason. The result is a sudden rush among hobby farmers to respond to the market interest. Sales of fresh turkeys at Christmas have doubled to 5 million over the past 10 years and they are still rising.

Wildlife Bonus: Twenty low-cost ideas for the living farm

The small-scale farmer can do a great deal to encourage wildlife in his paddocks and woodland – he has more space than most gardeners and faces fewer regulations than large-scale agricultural holdings.

If he is happy to operate outside the constraints imposed by grants and tax breaks, he can do lots of interesting things. Here are 20 low-cost things to do on your lifestyle farm which could make the place even better for the birds, the bees and the flowers.

1. Shrubs are vital for encouraging birds. The lawns and herbaceous borders of the town garden and the vegetable plots of the market garden are quite limited in their attractions when compared with a shrubbery. A useful species is *Berberis darwinii*, a familiar thorny-leaved evergreen from South America which produces golden flowers in May (when admittedly you may have overdosed on yellow daffodils and forsythia), followed in autumn by abundant berries. Darwin's barberry grows to 10ft in any soil and has the virtue of being particularly attractive to nesting longtailed tits, which favour this bush above all others. Readily grown from seed.

2. For half the year – the cold half – the easiest way to have interesting birds around the hobby farm is to feed them. The best food is wheat, or at least it's the most readily obtainable in quantity and it keeps for years in plastic dustbins. Pheasants, doves and waterfowl need little else. Invest in an automatic hopper which sprays wheat around, morning and evening, and the birds will love you. Cost is £146 from Solway Feeders (www.solway feeders.com) American feeders are even more sophisticated, powered by solar cells. Hunters use them to attract big bucks in the woods.

3. Rotary mowing of your larger lawns, allowing the cuttings to fall as a mulch, is the best way of encouraging blackbirds, thrushes, hedge sparrows, starlings and pied wagtails because the worm and insect population of the lawn will flourish. The downside: you get a strong growth of moss.

4. Sir David Attenborough has described butterflies as 'the canary in the coalmine' in the British countryside. Their early warning is quite clear: more than 70 per cent of butterfly species have declined in recent years. To give several species a boost in late summer plant buddleia and (unclipped) privet, and in the herbaceous border, the grey-leaved sedum Autumn Joy, whose flowers will be covered in glorious butterflies for weeks.

5. A curiosity worth trying, if you want to spot some interesting small birds during hot weather, is a drip puddle. A hose running very gently – a drip every 2–3 seconds – into a shallow basin in the shrubbery, preferably on the ground in the edge of woodland, will draw blackcaps and other warblers, quite fascinated by the splash. Allow the water to fall at least 2ft, to get the right noise and the 'plop' effect.

6. Wildflower meadows are not simple to establish, particularly if you have fertile soil where big weeds and trees will constantly try to take over, but they require no watering and no fertiliser. On poor soils, mow the meadow down to a three-inch stubble in autumn, allowing the cut stems to dry, so that seeds can fall to the ground. If you want to encourage spring-flowering meadow plants, mow in July.

7. Swap your tabby cat for a white one, which the birds can see coming. Or even better, a white cat with two blue eyes. She will probably be deaf (a hereditary handicap in 60–80 per cent of blue-eyed whites in Britain) and this cuts down her efficiency as a hunter by about 50 per cent. Domestic cats and fast traffic are the major killers of our birds and small mammals.

8. Establish some 'staging posts' for small birds in the more open parts of your hobby farm. In my own walled garden, which is 70 yards wide, there is a central fuchsia bush, nearly 10ft high, into which most tits and finches dive during the long crossing. This hardy *Fuchsia riccartonii* is densely twiggy and has saved many birds from the sparrowhawks.

9. The most rewarding project for birds and amphibians is a pond. Even a small and shallow pond will draw frogs, unless your smallholding is surrounded by frog-proof walls. Toads are more faithful to their ancient spawning sites. Make sure there's a shallow corner where froglets (and adventurous voles) can climb out.

10. Knock a window out of the stable and let the swallows in. And leave some dung to rot out in the paddock, encouraging the bugs on which the swallows feed. Two years ago, we sent 50 swallows off to Africa in the autumn and only three came back. However, we keep trying...

11. Your bird table will be safer for finches, tits and the small fry if it is sited close to evergreen shrubs and you might even try fencing it off with a box of 2.5 inch netting. This deters most predators and allows the pretty wee things to get to the food (although nothing stops grey squirrels).

12. Don't be shy about the fact that you are operating on a small-ish canvas. In total, British gardens and small-holdings exceed nature reserves in area by something approaching quarter of a million acres. Your 10 acres count.

13. Honey bees need all the help they can get – like butterflies, they find it a bit of a struggle in Britain's changing landscape. Early flowering trees and shrubs are valuable to nectar-seeking bees; try crab apples, cherries, horse chestnuts, berberis, cotoneaster, mountain currant and ceanothus. You don't need an excuse to plant any of these, but if asked, say you are doing it for the honey bees.

14. Dry dusting places are much appreciated by several kinds of birds, from pheasants to sparrows. A sheet of clear corrugated plastic raised on a 2ft frame is enough to provide a dry spot under a hedge or at the side of a wood.

15. One of the main advantages of working with 10 acres rather than 0.5 acre is that you can plant bigger, more robust things – Jerusalem artichokes being an example. They are just too rampant for the garden. However, they are brilliant on the smallholding. They grow strongly up to 6–10ft, providing a summer windbreak, and their knobbly roots are excellent food for pheasants. The birds will dig for them all winter and if they don't dig deep enough, a run through with a

Rotavator will reveal the tubers. Use some for your soup pot too.

16. Even the least skilled of us can make nest boxes, especially if you remember that the important finishing touch is a piece of waterproof roofing felt on top. There are two basic designs: the enclosed box and the half-open fronted box. Hole size for enclosed boxes is: tits, wrens, tree sparrows 29mm; house sparrows (becoming rarer, they need boxes too) 30mm; nuthatches 32mm; starlings, woodpeckers 5cm; jackdaws 15cm. Robins and spotted flycatchers use a half-open box about 10cm square. To attract a blackbird to a safe, secluded site, put up a half-front box about 20cm square. If you have a choice, a site facing east is recommended as cooler and usually dryer.

17. Bats can sometimes be tempted to use artificial roosting boxes. The basic design is a rough wood box with a waterproof roof, the bats finding their way in through a narrow slot (about 18mm wide) between the bottom of the box and the back wall. For ideas see the American website at www.backyardbird.com. Do not be discouraged by the story of the famous Florida tourist attraction, R.C.Perky's Bat Tower on Sugarloaf Key. This elaborate 30ft wooden tower was built to provide a roost for bats which would eat the islet's malarial mosquitoes. It has never attracted a single wild bat in more than 70 years.

18. Plant an oak tree in the paddock. This buzzes with bugs and is going to be enjoyed by your grandchildren, but is quite slow. To enjoy a beautiful shade tree in your own lifetime, plant a hybrid poplar.

19. Area not density is the key to success in planting small woods and shrubberies for birds. Spend the money on more fence rather than more plants.

20. *Note*: 'Wildlife' includes rats, moles and magpies; 'wildflowers' include ground elder, thistle and dock. Wildlife management is the science of encouraging the species you want and discouraging the rest.

Where to look for Further Information and where to buy the things mentioned in this book

Artichokes: Jerusalem artichokes from garden centres and also The Organic Gardening Catalogue (www.organicCatalog. com) 01932 253 666, approx 15 tubers of fuseau artichokes for £4.90 plus 80p carriage; Marshall's, Wisbech (www.marshalls-seeds.co.uk) 01945 583 407; W. Robinson and Sons, Sunny Bank, Forton, near Preston PR3 0BN (www.mammothonion.co.uk) 01524 791 210, 10 fuseau tubers for £7. The same sources have globe artichokes at about £4 each for spring-delivered plants, £1.20 for a packet of seeds.

Asparagus: The Organic Gardening Catalogue, Riverdene Business Park, Molesey Road, Hersham, Surrey KT12 4RG (01932 253 666): Asparagus crowns, several varieties, £23.95–£28.25 per 30; seed, £0.99p to £2.45 per packet. S.E. Marshall and Co. Ltd., Freepost NATE104, Wisbech, Cambridgeshire PE13 2BR (01945 466 711): Several varieties, crowns from £8.75 per 10 for March/April delivery. Asparagus Growers' Association, 133 Eastgate, Louth, Lincolnshire LN11 9QG (01507 602 427).

Apples: Suppliers of trees in pots, suitable for planting at any season, include Bernwode Plants, Kingswood Lane, Ludgershall, Bucks HP18 9RB (01844 237 415;

www.bernwodeplants.co.uk) Varieties stocked include Bulmer's Norman, Dabinett, Kingston Black, Sweet Coppin and many culinary/cider and dessert/cider types.

Other suppliers: Scotts Nurseries (Merriott) Ltd., Merriott, near Crewkerne, Somerset TA16 5PL (01460 72306), who have 200 varieties of apple; Cider Apple Trees, 12 Tallwood, Shepton Mallet, Somerset BA4 5QM (01749 34 33 68), who stock about 30 varieties of cider apple. Brogdale Horticultural Trust, Brogdale Road, Faversham, Kent ME13 8XZ (01795 535 286) www.brogdale.org.uk has the largest collection of apple varieties (more than 2200), selling and promoting fruit and welcoming visitors to the farm shop. There is a Cider Festival in September. Middle Farm, Firle, Lewes, East Sussex, has the National Collection of Cider and Perry (more than 250 ciders) and holds its Apple Festival in October. The farm is open 7 days, 10–5, and will press your own fruit in season. Call 01323 811 324/411 www.middlefarm.co.uk For a comprehensive list of UK cider makers, go to www.ciderandperry.co.uk This site also has suppliers of home-brew equipment and much else.

Beekeeping: British Beekeepers' Association, www.bbka.org.uk

Cobnuts: Blackmoor Fruit Nurseries, Blackmoor, Liss, Hampshire GU33 6BS. 01420 473576; Brogdale Horticultural Trust, Brogdale Rd, Faversham, Kent ME13 8XZ. 01795 535286; Deacon's Nursery, Moor View, Godshill, Isle of Wight PO38 3HW. 01983 840750; JIB Cannon & Son, Roughway Farm, Roughway Lane, Tonbridge Kent TN11 9SN. 01732 810260; S.E. Marshall & Co. Ltd., Wisbech, Cambs. PE13 2RF. 01945 583 407; RV Roger Ltd, The Nurseries, Pickering, N. Yorks, Y018 7HG. 01752 472226; Scotts Nurseries (Merriott) Ltd, Merriott, Somerset TA16 5PL. 01460 72306. Books: *Pruning Kentish Cobnuts*, by The Kentish Cobnut Association. (ISBN 0 9541605 0 9).

Management and pruning, varieties, pests and diseases. £3 plus P&P. *In a Nutshell – the story of Kentish Cobnuts,* by Meg Game. Available direct from the author at meg.game@london.gov.uk Meg Game is a professional ecologist who has researched the history of nut growing and studied the wildlife associated with the plats. She has a particular personal interest since she still lives on, and tends, the Kentish plat bought by her parents in 1939.

Donkeys and Mules: British Donkey Breed Society: 01732 864414 (www.donkeybreedsociety.co.uk) Miniature Mediterranean Donkey Association of UK: 01394 450615 British Mule Society: 01793 615478; (www.britishmule society.org.uk) Books: *The Professional Handbook of the Donkey,* Elisabeth D. Svendsen, £10.49; *Looking After a Donkey,* Dorothy Morris, £7.69; *The Mule,* Lorraine Travis, £29.95

Utility waterfowl: Information about restoring the egg-laying ability of utility breeds of waterfowl can be found on the website of TROUP (Time to Restore Our Utility Poultry) at www.utilitypoultry.co.uk – lots of background on traditional breeds of fowls and ducks. 01631 720 223; Domestic Waterfowl Club – Michael and Sylvia Hatcher, Limetree Cottage, Brightwalton, Newbury, Berkshire (01488 638 014) www.domestic-waterfowl.co.uk (breeds and breeders); British Waterfowl Association – Sue Schubert, PO Box 163, Oxted, RH8 0WP. For both domestic and ornamental (wild) waterfowl breeders. www.waterfowl.org.uk; Indian Runner Duck Association, set up in 2000 to bring together runner enthusiasts – Julian Burrell, 1 Coombe Cottages, Lamellion, Liskeard, Cornwall PL14 4JU www.runnerduck.net; www.poultrypages.com is useful starting point for breeds, breeders and UK auctions.

Blue eggs: Araucana Club of Great Britain – www.araucana. org.uk. Breeders of both Araucanas and Legbars include: Kintaline Farm, Argyll, 01631 720 223 (creamlegbar. co.uk); Mrs. L. Jackson, Flintshire, 01352 720 745; Ty-Celyn Poultry, N. Wales, 01352 780 005; Alison Taylor, Anglesey, 01407 839 119; Cream and Cotswold Legbars from 01386 858 007; Araucana breeders include: Pam's Poultry, Staffs, 01630 620 251; Invercharron Poultry, Sutherland, 01863 766 069; Charmaine Lovering, Hereford, 01981 240 766; All Pure Poultry, Norfolk, 01379 588 136; Ivan Mears, Bucks, 01525 261 606; David Bew, Wiltshire, 01225 754 706; Freightliners City Farm, North London, 0207 609 0467; Dee Terry, Hampshire, 07884 076 463; Fenton Poultry, Somerset, 01823 672 075; Bantams for the Garden, Kent, 01622 812 564; High Down Poultry, East Sussex, 01580 860 426. Breeders of Cream Legbars incl: Abington Poultry, Lincs, 01652 679 001; Delightful Chickens, Northants, 07876 342 256; Anulines, Essex, 01255 551 852; Holditch Poultry, Devon, 01460 221 602.

Game Crops: Seed suppliers include: Dunns (Long Sutton) Ltd: 01406 362 141; David Bright Ltd: 01722 712 722; Pearce Seeds: 01935 811 400; Advanta Seeds UK: 01529 304 511; Church of Bures: 01787 227 654; David Bell Seeds: 01968 678 480; Kings: 0800 58 79 797.

Goats: British Goat Society, Secretary, 34-36 Fore Street, Bovey Travey, Newton Abbot, Devon TQ13 9AD (01626 833 168) email secretary@allgoats.com website: www.allgoats.com Anglo Nubian Breed Society, Mrs M Edginton, 4, Wadehouse Lane, Drax Hales, Selby, North Yorkshire, YO8 8PN 01757 618756 www.anglo-nubian.org.uk British Alpine Breed Society, Mrs S. Head, The Old Tanyard, Pound Hill, Corsham, Wiltshire, SN13 9HT 01249 716350 www.britishalpines.co.uk. British

Saanen Breed Society, Mrs S Harcombe, The Field Stud Farm, Risbury, Leominster, Herefordshire, HR6 0NN. 01568 760649. British Toggenburgs, Mr Gordon Turner, Rouval, Ipswich Road, Langham, Colchester, Essex, CO4 5NG. 01206 230756 http://www.britishtoggenburgs.co.uk English Goat Breeders' Association, Mrs J Parry, Heathgate Farm, Gills Lane, Rooksbridge, Axbridge, Somerset, BS26 2TZ, 01934 750602, http://www.egba.org.uk/ Harness Goat Society, Mrs. E. Pratt, Honey Cottage, Pound Lane, Burley, near Ringwood, Hampshire BH24.4EB (also covers pack goats in the UK). Books: *Goats – A Guide to Management*, by Patricia Ross, Crowood Press, £9.99. *The Pack Goat*, by John Mionczynski, Pruett Publ., £13.99.

Grapes: Books – *Grapes: Indoors and Out*, by Harry Baker & Ray Waite, *RHS Wisley Handbooks*, published by Cassell, £7.99. *Successful Grape Growing for Eating and Winemaking*, by Read and Rowe, Groundnut Publishing, £12.95.

Vines: Sunnybank Vine Nurseries, Journey's End, Rowlestone, Ewyas Harold, Herefordshire HR2 0EE (01981 240 256) www.vinenursery.nerfirms.com

Hedges: planting information from UK Farming and Wildlife Advisory Groups at www.fwag.org.uk (with links to Scottish and Welsh sites). Recommended books include: *Hedging – A Practical Guide*, £13.95, from the British Trust for Conservation Volunteers (tel: 01302 572 244) www. btcv.org.uk and *New Hedges for the Countryside*, Murray Maclean (1992), which you may have to borrow from your library. Buckingham Nurseries' website at www.hedging. co.uk has photos of many types of hedge and can quote wholesale prices. Sample hedges are on view at the nurseries: Tingewick Road, Buckingham, MK18 4AE (01280 822133).

Grass seed and herbage mixtures: Cotswold Seeds, Moreton-in-Marsh, Gloucestershire GL56 0BF (0800.25 22 11) www.cotswoldseeds.com Grass Seed Direct (Ross Muirhead & Co.) Menstrie Mains, Menstrie, Clackmannanshire FK11 7AE (01259 760 400)www.grassseed.co.uk. British Seed Houses, Camp Road, Witham St. Hughes, Lincoln LN6 9QJ (01522 868 714) www.britishseed houses.com Countrywide Farmers, Bradford Road, Melksham, Wiltshire SN12 8LQ (01225 701 300) www.countrywide farmers.co.uk Barenbrug UK, Rougham Industrial Estate, Bury St. Edmunds, Suffolk IP30 9ND (01359 272 000) www.barenbrug.co.uk N. & F. Allan Seeds, Clinkerheel Drive, Birkhill, Angus DD2 5RN (0845 601 55 82) www.allanseeds.co.uk

Meat chickens: Meadowsweet Poultry have agents everywhere in the UK, selling utility birds and also 'Master Gris' pastured meat chickens. 0191 384 2259 (www.meadow sweetpoultry.co.uk). Neville Tilney, Greenacres Game Farm, Newton Road, Hainford, Norwich NR10 3LZ (01603 891092). Sasso and Ross meat hybrids from £55 per 100. Kintaline Farm, Benderloch, Oban, Argyll PA37 1QS. 01631 720223 (www.utilitypoultry.co.uk) Utility breeds. Cyril Bason, Bank House, Corvedale Road, Craven Arms, Shropshire SY7 9NG (01588 673 204) www.cyril-bason.co.uk Ross/Cobb meat chicks from £60 per 100; deliveries over most of England and Wales.

Pigs: Secretaries of the seven British rare breeds of pig can be contacted via the Rare Breeds Survival Trust, National Agricultural Centre, Stoneleigh Park, Warwickshire CV8 2LG (tel: 024 7669 6551, fax: 024 7669 6706; email: enquiries@rbst.org.uk website: www.rbst.org.uk Book: *Rare Breed Pig Keeping*, edited by Richard Lutwyche, £4.95 plus 80p p&p from Marketability, Freepost (GL442),

Cirencester, Glos. GL7 5BR (tel: 01285 860 229); also *The Pigman's Handbook*, £15 from RBST. Oxford Sandy and Black Pig Society, Hon. Secretary Mrs. Cathie Annetts, Tadneys Farm, Fox Lane, Kempsey, Worcester WR5 3QD (tel: 01905 821 828).

Poplars: The Poplar Tree Company – www.poplartree.co.uk Bowhayes Farm – www.bowhayesfarm.co.uk Brownfield Remediation to Forestry Research Group, John Moores University – http://cwis.livjm.ac.uk

Potatoes: GM & EA Innes, Potato specialists – 71 varieties (plus 24 heritage varieties from the MacLean Collection), Oldtown, Newmachar, Aberdeenshire AB21 7PR, 01651 862333. Edwin Tucker and Sons Ltd, Brewery Meadow, Stonepark, Ashburton, Newton Abbot, Devon TQ13 7DG, 01364 652403.

Poultry: Poultry Club of Great Britain, 30 Grosvenor Road, Frampton, Boston, Lincolnshire PE20 1DB. www.poultryclub.org Links to all the breed clubs, plus useful advice pages. Suppliers of equipment and stock: Domestic Fowl Trust, Station Road, near Evesham, Worcestershire WR11 7QZ (01386 833083) www.domesticfowltrust.co.uk; Kintaline Poultry Centre, Benderloch, Oban, Argyll PA37 1QS (01631 720 223) www.kintaline.co.uk; Cass's Chucks, Lancashire 01706 812 710; Glencroft Poultry, North Yorkshire 01748 832 786; Ty-Celyn Poultry, North Wales, 01 352 780 005; Robert Jefferies, Norfolk 01603 754 390; Fine Feathers, Staffordshire 01782 504 460; Wernlas Collection, Ludlow, Shropshire, 01584 856 318, www.wernlas.com; Cotswold Farm Park Poultry, Gloucester, 01451 860 322; Thorne's Poultry Centre, Herts., 01462 675 767; Ascott Smallholding Supplies, Dudleston Heath, Ellesmere SY12 9LJ www.ascott.biz 0845 130 6285; Forsham Cottage Arks, Goreside Farm, Great

Chart, Ashford, Kent TN26 1JU 0800 163 797
wwwforshamcottagearks.com (agents throughout UK);
Black Rocks: call 01577 840 401 for details of local
suppliers of pullets. Bovans Nera hybrids, Meadowsweet
Poultry, 0845 165 1532, www.meadow sweetpoultry.co.uk.
Books: Numerous titles covering almost all large breeds and
bantams, from Beech Publishing House, 7 Station Yard,
Elsted Marsh, Midhurst, West Sussex GU29 0JT 01 730 810
524. Other titles: *Starting with Chickens*, £6.95; *Chickens at
Home*, £6.50; *Modern Free Range*, £6.50; *Practical Poultry
Keeping*, £14.99; *Hens in the Garden*, £7.99; *Poultry for
Anyone*, £19.99; *Garden Poultry Keeping*, £9.50.

Rabbits: British Rabbit Council, Purefoy House, 7 Kirkgate,
Newark, Nottinghamshire NG24 1AD (01636 676 042).
www.thebrc.org; booklet: *Getting Started in Rabbits*,
£1.50. Stock and equipment: Eaudike Commercial Rabbits,
Northamptonshire (01775 840 691) www.eaudike rabbits.
co.uk. Processors: Woldsway Foods Ltd., Spilsby, Lincolnshire PE23 5RG (0800 298 5000) www. woldsway.co.uk .

Rhubarb: Pot-grown rhubarb varieties are on sale all year
round from garden centres. Varieties on sale by mail order
for winter delivery include: Victoria, popular April-June
variety, and Glaskin's Perpetual, red-tinged green stems,
both £2.49, from Buckingham Nurseries, Tingewick Road,
Buckingham MK18 4AE (01280 822 133 www.hedging.
co.uk); Stockbridge Arrow, modern forcing variety, £9.95,
and Victoria, £8.45, both for packs of 3 crowns, from S.E.
Marshall & Co., Freepost Nate 104, Wisbech,
Cambridgeshire PE13 2BR (01945 583 407); Stockbridge
Arrow, £9.75 for 3 crowns from D.T. Brown, Bury Road,
Newmarket, Suffolk CB8 7PR (0845 166 22 75 www.
dtbrownseeds.co.uk); Victoria and Champagne, early
scarlet, both £8.45 for 3 crowns, also seed of Glaskin's

Perpetual, £1.24 per packet from the Organic Catalogue (0845 130 1304 www.organicCatalogue.com). Wakefield Rhubarb Festival is held in early February each year. Details from Wakefield Tourism (01924 305 911).

Sheep: Rare Breeds Survival Trust www.rbst.org.uk; National Sheep Association www.nationalsheep.org.uk; Cotswold Farm Park (open April-Sept) www.cotswoldfarm park.co.uk; Association of Guilds of Weavers, Spinners and Dyers www.wsd.org.uk; Hebridean Woolhouse www.hebridean woolhouse.com; Cotswold Woollen Weavers 01367 860 491, email wool.weavers@dial.pipex. com; Association of Guilds of Weavers, Spinners and Dyers (website lists scores of local guilds throughout the UK) www.wsd.org.uk. Johnstons of Elgin, Newmill, Elgin, Moray IV30 4BR, 01343 554000 www.johnstonscashmere.com

Soft Fruit: Pack of 6 Kilner-type preserving jars £10.56 from Ascott Smallholding Supplies, Dudleston Heath, Ellesmere, Shropshire SY12 9LJ, 0845 130 6285. Clip-top and screw-top preserving jars also from Wares of Knutsford, 36A Princess Street, Knutsford, Cheshire (01565 751 477; www.waresofknutsford.co.uk); Kitchenary Cookshop, Taverham, Norwich NR8.6HT (01603 261 932; www.zednet.co.uk/kit/) and most department stores.

Turkeys: Turkey Club UK: www.turkeyclub.org.uk Heal Farm, King's Nympton, Devon: www.healfarm.co.uk Pipers Farm, Cullompton, Devon : www.pipersfarm.com Rookery Farm, Thuxton, Norfolk (breeders of Norfolk blacks since 1880): 01362 850 237. Kelly Turkey Farms, Springate Farim, Danbury, Essex: www.kelly-turkeys.com